STORMWATER DETENTION

For Drainage, Water Quality, and CSO Management

Peter Stahre

Malmö Water and Sewer Works
Malmö, Sweden

Ben Urbonas

Urban Drainage and Flood Control District
Denver, Colorado

Prentice Hall
Englewood Cliffs, New Jersey 07632

Library of Congress Cataloging-in-Publication Data

Stahre, Peter.
    Stormwater detention : for drainage, water quality, and CSO
management / by Peter Stahre and Ben Urbonas.
        p.   cm.
    Includes bibliographies and index.
    ISBN 0-13-849837-7
    1. Storm water retention basins. 2. Urban runoff.   I. Urbonas,
Ben. II. Title.
TD665.S73 1990
628'.21--dc20   89-33449
                                                          CIP

Editorial/production supervision and
  interior design: **Brendan M. Stewart**
Cover design: **Lundgren Graphics, Ltd.**
Manufacturing buyer: **Mary Ann Gloriande**

This book can be made available to businesses
and organizations at a special discount when
ordered in large quantities. For more information
contact:

Prentice-Hall, Inc.
Special Sales and Markets
College Division
Englewood Cliffs, N.J. 07632

© 1990 by Prentice-Hall, Inc.
A Division of Simon & Schuster
Englewood Cliffs, New Jersey 07632

Printed in the United States of America
10 9 8 7 6 5 4 3 2

ISBN 0-13-849837-7

Prentice-Hall International (UK) Limited, *London*
Prentice-Hall of Australia Pty. Limited, *Sydney*
Prentice-Hall Canada Inc., *Toronto*
Prentice-Hall Hispanoamericana, S.A., *Mexico*
Prentice-Hall of India Private Limited, *New Delhi*
Prentice-Hall of Japan, Inc., *Tokyo*
Simon & Schuster Asia Pte. Ltd., *Singapore*
Editora Prentice-Hall do Brasil, Ltda., *Rio de Janeiro*

To our wives and daughters

	Irena
Malina	Tesa
Misia	Vida
—P.S.	—B.U.

Table of Contents

Foreword

In the last 25 years, stormwater management has become an important part of the general field of water resources engineering and management. Today it has become clear that it is an important subject that must be addressed to solve problems of integration in water management. Stormwater is the integrating factor; it connects water sources to receiving waters. Moreover, stormwater provides amenities in urban areas when it is managed with open space and recreation. Stormwater management can help to achieve numerous objectives of urban development, including: flood control, water quality enhancement, and generally better living conditions in urban areas.

Serious interest in stormwater management began in the U.S. in the 1960s, perhaps as a result of the adverse consequences of the urbanization that followed World War II. In the 1970s three concerns emerged: stormwater quality, the development of mathematical models to simulate stormwater, and the use of detention storage.

It turned out that all of these new subjects shared two characteristics: They were important to the field, and they were complex. To deal with the complexity, a definitive and detailed text on detention storage has been needed for a long time, and with the publication of this text we finally have one.

It is appropriate that an international team prepared the book: Ben Urbonas and Peter Stahre. They have collaborated in the field for some years, mostly through activities organized by the American Society of Civil Engineers' Urban Water Resources Research Council. Ben Urbonas was its chairman for two years and continues to be one of its most faithful participants. Ben also organized for the Council the Engineering Foundation Conferences on detention storage and on urban stormwater quality. At the same time, Peter Stahre published extensively on these topics in Europe, including a Swedish language book on stormwater detention.

Detention storage planning and design may be the most important technical subject in the field of stormwater management. As pointed out by General William Whipple in some of his writing, it has dual purposes: detention of flood waters and treatment of the quality of stormwaters. Detention is one of the subtle tools of stormwater planning, design, and operations. It is attractive on first impression, but may even aggravate the downstream conditions if done incorrectly. The systems may not work if they are not designed carefully, and they can become inoperative if not maintained.

No treatment of the subject would be complete without mentioning the contributions of the Urban Water Resources Research Council, and in particular its program leader for many years, Murray B. McPherson. "Mac" was the leader in the field of stormwater management. He thought about it in an integrated manner. The Council's publications under his direction are excellent and comprehensive. A review of these does not turn up any on the single subject of detention storage, but they motivated the ASCE-organized Engineering Foundation Conference and were very influential on the development of theory and practice in the field.

The Council's publications are listed in an appendix of my book *Urban Water Infrastructure: Planning, Management and Operations,* John Wiley & Sons, 1986. As far as I know, this is the only commercially available list of the Council's publications under Mac's leadership. ASCE still may be able to furnish lists, but they do not distribute the publications. Most have National Technical Information Service (NTIS) numbers and can be ordered from NTIS.

Other members of the Urban Water Resources Research Council deserve credit for contributions and for encouragement over the years. They cannot all be mentioned here, but many of them appear in the citations of the text. Scott Tucker's contributions must, of course, be mentioned, both from his early service with Murray McPherson as deputy director of the research program and later as Director of Denver's Urban Drainage and Flood Control District, where Ben is chief of its planning program.

I congratulate Ben Urbonas and Peter Stahre for compiling this excellent presentation of the theory and practice of using detention storage in stormwater management. Their contribution will certainly be of great value to the profession for many years.

Neil S. Grigg
Department of Civil Engineering
Colorado State University
Fort Collins, Colorado

Preface

Urban stormwater engineering and management has advanced more in the last 20 years than at any time in history. Granted, the advances were possible only because the basic mathematic, hydraulic, and hydrologic principles were there to act as a foundation on which many individuals and institutions built this technology. As a part of these advances, stormwater detention emerged as one of the basic components used today in stormwater management. When properly applied, it can reduce the adverse impacts of accelerated stormwater runoff from urbanized lands by slowing or disposing of the runoff and by removing from urban runoff some of the pollutants that it is known to carry.

This book was written with the practicing engineer and stormwater manager in mind. Practitioners should find it a useful reference. At the same time, it contains sufficient materials to serve as a textbook for a college or training course in urban stormwater engineering.

Many have contributed to the technology described in this book. The authors sincerely hope that adequate acknowledgment has been given to all contributors and sources of information. The authors particularly wish to acknowledge the Swedish Council for Building Research for its financial support of basic research that led to the compilation of much of the material contained in this book.

—Peter Stahre
—Ben Urbonas

1

Overview

1.1 DEFINITION OF STORAGE FACILITIES

The term *storage facilities* is used in this text to describe any combination or arrangement of detention and retention facilities in a combined sanitary–storm sewer system or a separate stormwater conveyance system. Unfortunately, there is no standard nomenclature to describe various types of storage facilities in this field of engineering. Therefore, it is up to each author to describe his or her terminology.

A classification for storage facilities is defined here which will be used throughout this text. It is based on the location of the facility in the sewer or flow conveyance system. This nomenclature is intended to help the reader to understand the different storage systems discussed in this book.

1.2 SOURCE CONTROL VS. DOWNSTREAM CONTROL

Since the basis for classifying storage facilities is the location in the collection and conveyance system, storage facilities can be classified as *source control* or *downstream control*. In the first case, the storage takes place far up in the wastewater or storm runoff collection system. Each storage facility is small and is located near the source, thereby permitting more efficient utilization of the downstream conveyance system. Some generalized observations can be made for source control:

- It affords great flexibility in choosing sites for facilities.
- Storage unit design can be standardized.
- Flow efficiency of the existing downstream conveyance system can be increased.
- Real time flow control can increase system capacity.
- It is difficult to monitor large numbers of storage units for proper design, installation, and upkeep.
- Maintenance and operation costs can be high due to large numbers of storage units.

With downstream control, on the other hand, the storage volume is consolidated at fewer locations. The storage facilities can be, for example, located at the downstream end of a large watershed, a subbasin of the watershed, or at a wastewater treatment plant. The following generalized observations can be made regarding downstream control:

- It has reduced construction cost as compared with a large number of source control facilities (Hartigan, 1982, 1986; Wiegand et al., 1986).
- It has reduced maintenance and operation costs.
- It is easier to administer construction and upkeep.
- Finding acceptable sites can be difficult.
- Land acquisition costs can be high.
- In combined sewer systems, fitting downstream storage into the sewer system can be difficult.
- Large dams and/or storage facilities can encounter public opposition.

There is no clear boundary of what constitutes source control and downstream control. There are storage facilities that, strictly speaking, can be classified as either of the two types. To further clarify the terminology, the classifications described earlier can be further refined. The more detailed description contains six categories of storage facilities shown as a block diagram in Figure 1.1.

1.3 LOCAL DISPOSAL

Local disposal of stormwater has gained considerable recognition and acceptance in recent years. Some communities in the United States, such as the states of Maryland and Florida, have mandated its use in new land develop-

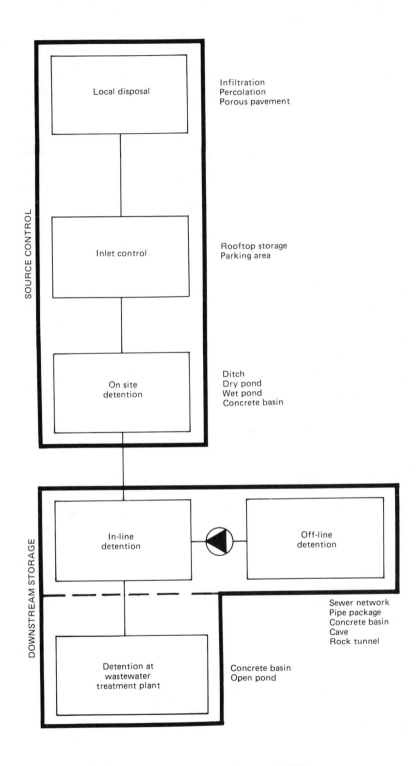

Figure 1.1 Classification of storage facilities.

ment. The term *local disposal* is used to describe storage facilities that use infiltration or percolation to dispose of stormwater. This practice attempts to utilize nature's own way of disposing of stormwater for the smaller storm events.

Where soil is suitable, stormwater from impervious areas is conveyed to an acceptable site covered with vegetation and infiltrates into the ground. If adequate infiltration sites are not available, stormwater can be routed to underground storage vaults from which water is allowed to percolate into the ground. Chapter 2 describes the principles and technology of local disposal.

1.4 INLET CONTROL AT SOURCE

Stormwater can be controlled at the source by detaining it where precipitation falls. This is done by choking off the inlets to the conveyance system. The detention volume is then obtained on properly prepared

- rooftops,
- parking lots,
- industrial yards, or
- other properly designed surfaces.

Chapter 3 describes in greater detail inlet control facilities.

1.5 ON-SITE DETENTION

On-site detention identifies an entire plethora of storage facilities which occur in the upper reaches of the flow conveyance system. The primary difference between on-site detention and local disposal and inlet control facilities is in the amount of tributary area being intercepted. On-site detention generally intercepts runoff from several pieces of real estate or from an entire subdivision. This means the water has been conveyed at least a short distance before it arrives at the detention facility. On-site detention may take any of the following forms:

- swales or ditches,
- dry basins (i.e., dry ponds),
- wet ponds (i.e., ponds with permanent water pool),
- concrete basins, usually underground, and
- underground pipe packages or clusters.

On-site detention ditches and ponds are described in Chapter 4, and concrete basins in Chapter 5.

1.6 IN-LINE DETENTION

The term *in-line* refers to detention storage in sewer lines, storage vaults, or other storage facilities that are connected in-line to the conveyance network. In-line detention can utilize excess capacity that may be found in an existing sewer network, or it may be necessary to construct separate storage facilities to provide the needed volume. In-line detention storage may take on any of the following forms:

- concrete basins,
- excess volume in the sewer system,
- pipe packages,
- tunnels,
- underground caverns, or
- surface ponds.

Chapter 6 describes the use of existing sewer systems, while Chapters 4, 5, 7, and 8 describe open ponds, concrete basins, pipe packages, and tunnel storage, respectively.

1.7 OFF-LINE STORAGE

Off-line refers to storage that is not in-line to the sewer network or other conveyance systems. Off-line storage is achieved by diverting the flow from the conveyance system to storage when a predetermined flow rate is exceeded. The diverted water is stored until sufficient conveyance or treatment capacity becomes available downstream.

With off-line storage, one must decide how the stored volume will be emptied. In designing off-line facilities, the following must be considered:

- the holding time needed to avoid odor or public health problems;
- hydraulic capacity or the treatment capacity of the downstream system;
- the hydraulic load on the downstream system at any given time; and
- the possibility for additional inflow before the storage is emptied.

1.8 STORAGE AT TREATMENT FACILITIES

Storage at treatment facilities is a special case of in-line or off-line storage. It is designed to provide flow or waste load equalization at the treatment plant. For an in-line storage, all flow passes through it, while in an off-line storage facility the flows that exceed a preset flow rate are stored. Both types of storage are usually emptied using pumps. Sometimes detention at the treatment plant can occur in the oversized sedimentation or aeration tanks of the treatment plant. Flow equalization at a wastewater treatment plant is discussed in more detail in Chapter 9.

REFERENCES

HARTIGAN, J. P., "Regional BMP Master Plans," *Urban Runoff Quality—Impact and Quality Enhancement Technology,* American Society of Civil Engineers, New York, 1986.

HARTIGAN, J. P., AND QUASEBARTH, T. F., "Urban Nonpoint Pollution Management for Water Supply Protection: Regional vs. Onsite BMP Plans," *Proceedings of Twelfth International Symposium on Urban Hydrology, Hydraulics, and Sediment Control,* University of Kentucky, Lexington, Ky., 1985.

WIEGAND, C., SCHUELER, T., CHITTENDEN, W., AND JELLICK, D., "Cost of Urban Quality Controls," *Urban Runoff Quality—Impact and Quality Enhancement Technology,* American Society of Civil Engineers, New York, 1986.

2

Local Disposal by Infiltration and Percolation

2.1 GENERAL

The traditional method of disposing of stormwater in an urban area is to drain it away as rapidly as possible. This is done by swale, gutter, and storm sewer conveying runoff to the nearest stream or river. In recent years, however, environmental concerns have arisen, and we are beginning to question the impacts on the receiving waters of continuing to use the practice of rapid conveyance downstream.

Williams (1982) and others have reported that the traditional approach to drainage increased flooding and stream erosion. In addition, questions are being asked regarding how the traditional approach affects the natural water balance, causes adverse shock loads of pollutants in the receiving waters, or, in the case of combined sewers, contributes to treatment plant malfunction.

In response to various concerns, some communities have elected to encourage or mandate local disposal of stormwater at its source of runoff. This is done by having a portion of the stormwater infiltrate or percolate into the soil. Although this approach has received greater attention in recent years, it has, in fact, been in use for a long time. Typically, it has been used to control stormwater from individual residential lots. The advantages often cited for the use of local disposal include:

1. recharge of groundwater;
2. reduction in the settlement of the surface in areas of groundwater depletion;
3. preservation and/or enhancement of natural vegetation;
4. reduction of pollution transported to the receiving waters;
5. reduction of downstream flow peaks;
6. reduction of basement flooding in combined sewer systems; and
7. smaller storm sewers at a lesser cost.

Equally good arguments are made against the use of local disposal systems. As a result, the use of local disposal systems should be addressed on a site-specific basis at each urban community. Arguments against the use of these facilities include:

1. The majority of runoff occurs from streets and large commercial areas, and local disposal of residential lots may have little impact on runoff.
2. Soils may seal with time, leaving property owners with a failed disposal system.
3. Very large numbers of infiltration and percolation facilities may not receive proper maintenance.
4. Reliance on their operation may leave communities facing enormous capital costs in the future, if or when these systems begin to fail.
5. Groundwater level may rise and cause basement flooding or damage to building foundations.

2.2 PRECONDITIONS FOR LOCAL DISPOSAL

The ability of soil to absorb stormwater depends on the following factors, among others:

- vegetative cover,
- soil type and conditions,
- groundwater conditions, and
- quality of stormwater.

Whenever precipitation falls to the ground, some of the water will infiltrate into the soil. The process of downward movement of the infiltrated water through the unsaturated zone above the groundwater table is called *percolation*. Further movement through the ground is called *groundwater flow*.

The upper layers of soil are strongly influenced by frost, plant roots, and other external factors. Therefore, the upper soil layers can be loosened by the external factors and be relatively porous. In clay soils, the upper layers may have some porosity in the form of cracks and cavities, but the lower layers are impermeable. As a result, areas containing substantial percentages of clays are poor candidates for local disposal through the use of infiltration or percolation. Unfortunately, the technology of water transport in soils is beyond the scope of this book, and the reader is referred to literature on geohydrology.

2.2.1 Vegetation

Figure 2.1 schematically depicts the exchange of water in the vegetation cover–soil complex. Much of the water that infiltrates into the soil is absorbed by plant roots and is returned to the atmosphere through plant respiration.

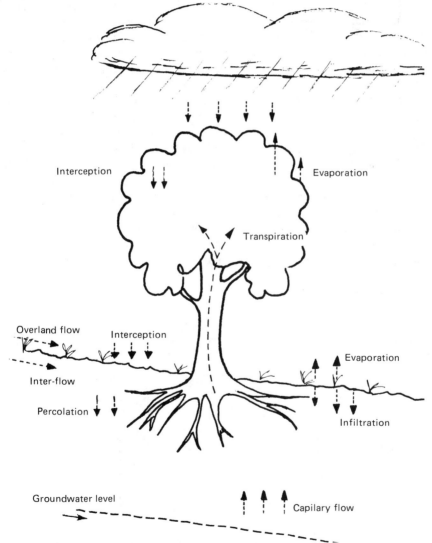

Figure 2.1   Water exchange in vegetation-covered soil.

The process of returning water to the atmosphere through plant respiration and through direct evaporation is called *evapotranspiration.*

The soil–vegetation complex functions, to a certain degree, as a filter which reduces clogging of the surface pores of the soil. That roots help keep the soil pores open and facilitate the buildup of the organic soil humus layer may explain why infiltration capacities are found to be higher in older grass-covered areas and in areas of least disturbance by human activities.

2.2.2 Soil Conditions

The quantity of water that can infiltrate into soil is, to a large extent, dependent on the effective porosity of the soil. *Effective porosity* is defined as the quantity of water that can be drained out from saturated soil. The capillary-bound water is therefore not included in the effective porosity definition. Table 2.1 summarizes the approximate values of porosity for different types of soils.

Permeability is a measure of how fast water can move through a soil. Table 2.2 gives the ranges of permeability typically found in various soil groups. By knowing the permeability and the slope of the groundwater's surface, the velocity of groundwater flow can be calculated.

Soils are never truly homogeneous, and field data is needed to effectively estimate the infiltration capacities at any site. To obtain reliable estimates of infiltration, the engineer needs to know the type of soils, the vertical thickness and horizontal distribution of each type, the presence of clay or other impervious lenses, and information about groundwater. This type of information can only be obtained by a drilling and sampling program that also includes field infiltration and percolation tests and observations of groundwater levels.

2.2.3 Groundwater Conditions

In addition to soils information, the engineer must also understand the groundwater conditions at a potential disposal site to determine the site's suit-

TABLE 2.1   Approximate Values of Soil Porosity

Type of Soil	Percent Effective Porosity
Crushed rock	30
Gravel and macadam	40
Gravel (2 to 20 mm)	30
Sand	25
Pit run natural gravel	15–25
Till (boulder clay)	5–10
Dry crust clay	2–5
Clay and silt (below surface)	0

ability. Among others, the following data are needed to understand the local groundwater conditions:

- distance between the surface of the ground and the groundwater;
- the slope of the groundwater surface;
- depth and direction of groundwater flow including zones of surface inflow and outflow; and
- variation in groundwater levels with the season.

In relatively homogeneous soils, the groundwater level essentially follows the slope of the land. Runoff patterns develop, however, as the terrain varies in slope, and regions of inflow and outflow become apparent, as illustrated in Figure 2.2. For obvious reasons, infiltration facilities and percolation facilities have to be located in the inflow regions. Attempts to infiltrate in the outflow regions will generally fail.

2.3 ARRANGEMENTS FOR LOCAL DISPOSAL

Local, site-specific conditions often dictate the selection of what type of local disposal facilities are used. The following sections describe how various infiltration and percolation principles need to be considered in the selection of

TABLE 2.2 Permeability of Various Soils

Type of Soil	Range in Permeability (m/year)
Gravel	30,000 to 3,000,000
Sand	30 to 300,000
Silt	0.03 to 300
Boulder clay	0.003 to 30
Clay	Less than 0.03

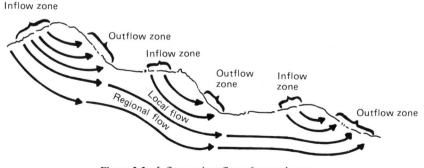

Figure 2.2   Inflow and outflow of groundwater.

such facilities. Examples are used to describe some of the more commonly observed facilities.

Before various types of installations are described, we discuss in general the selection of the type of local disposal. Each type will depend on the goals for local disposal and will dictate how the disposal facility is designed. The variety of disposal modes can be illustrated through the use of a conceptual, so-called geometric model, shown in Figure 2.3. Because of its conceptual construction, it also describes possibilities that lack practical significance.

The conceptual model consists of a number of schematic stormwater hydrographs located between two concentric circles. The various local disposal concepts are characterized by the radii of the circles intersecting the hydrographs. Any disposal concept can then be described with the aid of the central angle, α. Four local disposal concepts (i.e., A,B,C, and D) are described in more detail to illustrate this conceptual model.

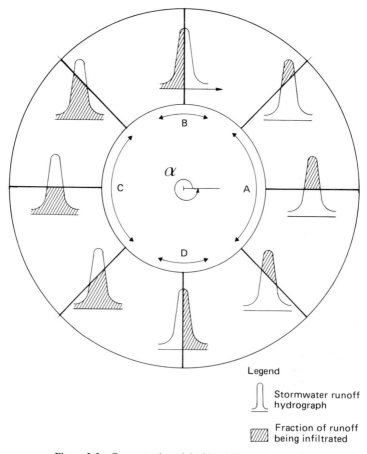

Figure 2.3   Conceptual model of local disposal concepts.

Concept A.   Only the flow peaks of stormwater runoff are intercepted and disposed locally, while the lesser flows are carried away by the storm conveyance system. To achieve this, the storm conveyance system is designed to carry away flows up to a certain design capacity, and the excess is diverted to the local, off-line disposal facility.

The advantage of this approach is that the size of the downstream storm conveyance system can be reduced (i.e., smaller storm sewers, culverts, etc.). On the other hand, the sometimes most strongly polluted fraction of runoff is carried away instead of being disposed through infiltration or percolation.

Concept B.   The intent of this concept is to intercept and to dispose locally the front end of the stormwater runoff hydrograph. This can be accomplished by routing all runoff into an interception facility until a predetermined volume is reached, at which time the remainder of the runoff continues downstream.

The advantage of this concept is that the sometimes most polluted initial storm runoff is intercepted and disposed through infiltration or percolation. The interception volume can be made large enough to intercept many, but not necessarily all, the storm runoff events. The percentage of the runoff events that are intercepted completely depends on the volume of storage and how that volume compares to the statistical distribution of the runoff events at the site.

Concept C.   In this concept, the runoff is diverted to the disposal facility up to a preset flow rate which cannot be exceeded. This can be accomplished by routing runoff through a small diameter pipe to the disposal facility. When the pipes flow capacity or the interception volume is reached, the larger flows overflow to the downstream conveyance system.

The advantages of this concept are almost identical to the aforementioned Concept B advantages. This design approach will also intercept the entire runoff from smaller storm events. The percentage of runoff intercepted depends not only on the volume of runoff, as in Concept B, but is also controlled by the preset maximum inflow rate. As a result, the runoff from some of the smaller storms may not be intercepted completely if the runoff hydrograph has a high peak, which is possible with intense, short-duration rainstorms.

Concept D.   This concept probably has the least practicality for local disposal. It is based on the interception of stormwater runoff after a preset runoff volume has occurred or a preset amount of time has elapsed. On the other hand, this concept may have utility in a real time control of a storm sewer system where the storm sewer capacity of the downstream watershed is optimized by rapidly conveying away the initial runoff before the flow from upstream tributary areas arrives.

An analysis of local goals, objectives, and system constraints will dictate which of the aforementioned concepts best fits the needs of the site. The selection of any one of the local disposal concepts cannot be made without first understanding local needs, rules, and regulations and how the concept will fit into the total stormwater management system of the community.

2.3.1 Types of Installations

The most commonly used local disposal installations include:

- infiltration beds,
- open ditches,
- infiltration ponds,
- percolation basins,
- pipe trenches, and
- rock-filled trenches or pits.

2.3.2 Infiltration Beds

The simplest form of local disposal is simply to let stormwater run onto a vegetation-covered surface, as illustrated in Figure 2.4. In areas with high groundwater table or fine-grained soils, infiltration can be very slow and will result in standing water. Under these conditions, it may be necessary to install subdrains, which are connected to the downstream conveyance system. This will return runoff to the conveyance system; however, its rate of flow will be significantly reduced and many of the impurities will be filtered out by the passage of the water through the soil.

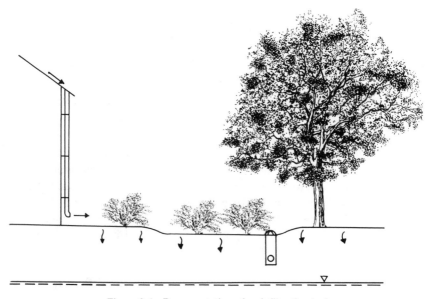

Figure 2.4   Representation of an infiltration bed.

2.3.3 Open Ditches

The use of ditches adjacent to roads or small parking lots, as illustrated in Figure 2.5, is a special case of surface infiltration. The rate of infiltration and the capacity for infiltration will depend on groundwater level, porosity of the soil, suspended solids load in stormwater, and the density of vegetation on the surface.

Good vegetation growth is essential. It can reopen and reaerate soil surfaces that become clogged. Eventually, as the surface soils accumulate sufficient sediments and infiltration rates are reduced, it will be necessary to excavate and dispose of the deposited fine sediments in order to restore infiltration capacity.

2.3.4 Infiltration Ponds

Stormwater detention or retention ponds located in permeable soils may also be used as infiltration ponds. This concept is shown in Figure 2.6. By their nature, ponds generally have a smaller surface area, in proportion to the tributary area, than other infiltration concepts described earlier. The larger surface loading rates can cause standing water to be present for extended periods of time. Healthy vegetative growth is therefore rarely possible in infiltration ponds. As a result, the infiltrating surfaces have a tendency to plug quickly. Once plugged, runoff will spill downstream and may overload the downstream conveyance system.

Ponds, as local disposal facilities, are less attractive than the other concepts described so far. Because of their tendency to rapidly lose infiltration capacity with time, they often become a nuisance. There are reported cases of property owners regrading their land to fill such ponds in the hope of eliminating drainage problems on their land. This often is done without realizing that their actions may be adding to drainage problems downstream.

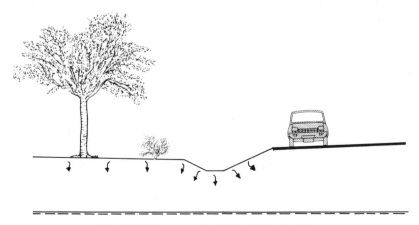

Figure 2.5   Representation of an infiltration ditch.

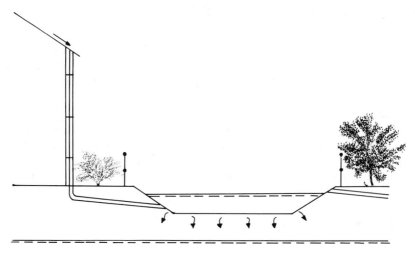

Figure 2.6   Representation of an infiltration pond.

Preventive maintenance to remove silt buildup at infiltration ponds can reduce these problems. On the other hand, this maintenance may have to be forced onto the property owners by a local government, which is politically unpopular. Thus, on-site infiltration ponds are generally destined to fail within relatively short period of time after installation.

2.3.5 Percolation Basins

The use of percolation basins for stormwater disposal was promoted in the early 1970s. A percolation basin is constructed by excavating a pit, filling it with gravel or crushed stone, and then backfilling over the top of the rock. The rock media provide the porosity for temporary storage of water so that it can then slowly percolate into the ground. One example of this for an installation at a single residence is illustrated in Figure 2.7.

Obviously, percolation basins are only feasible if they are located in suitable soils and groundwater conditions. If the basin is located in the unsaturated soil zone, namely, above the groundwater, stormwater will percolate out of the basin and raise the level of the groundwater. Basins should not be designed to be completely or partially submerged by groundwater, as they will not function properly. If the permeability of the surrounding soil is low, it will be necessary to install a bottom drain pipe similar to the one illustrated in Figure 2.8, which is connected to the downstream conveyance system via a choked outlet.

A serious problem with percolation basins is their tendency to clog with suspended particles carried by stormwater. Once clogged, all the rock needs

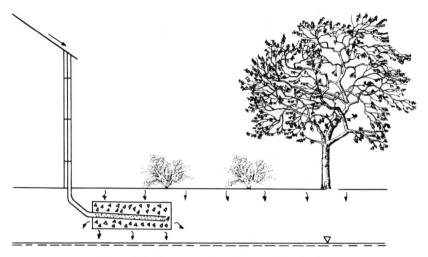

Figure 2.7   Representation of a percolation basin.

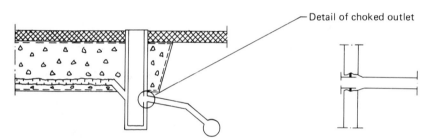

Figure 2.8   Choked bottom outlet of a percolation basin. (After Swedish Association of Water and Sewer Works, 1983.)

to be removed and replaced with clean rock, and the pit itself needs to be enlarged to remove clogged soils adjacent to the rock. This problem can be significantly reduced if the stormwater is filtered before it enters the rock-filled basin.

Filtering can be accomplished by passing the stormwater through a granular filter bed or a fine mesh geotextile material. In either case, the filter media will need to be removed and replaced from time to time, but at considerably less cost than the replacement of the rock bed. Figure 2.9 illustrates a filter installation upstream of a percolation basin.

Another problem that needs to be addressed in constructing percolation basins is the natural phenomenon of finer soil particles migrating into the pores of the coarse rock media. This can be prevented through the use of properly selected geotextile fabrics or the installation of granular filtering layers between the rock media and the adjacent in situ soil.

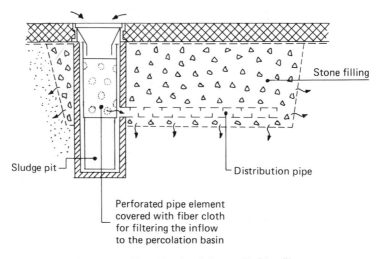

Figure 2.9 Example of an inflow well with a filter.

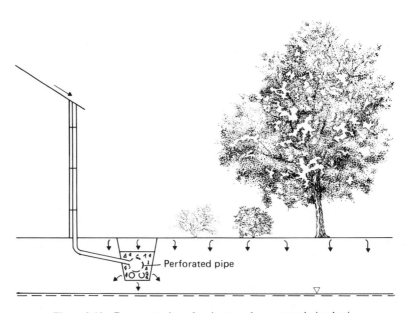

Figure 2.10 Representation of a pipe trench as a percolation basin.

2.3.6 Pipe Trench

Pipe trench is nothing more than a special case of percolation basin. A sample installation is shown in Figure 2.10, which illustrates how stormwater is stored not only in the rock media, but in the pipe itself. All the problems associated with clogging of rock media in the percolation basin can be found in the pipe trench. Therefore, the filtering of stormwater and the prevention of soil migration into the rock media must be provided for a successful pipe trench installation.

2.4 SPECIAL CONSIDERATIONS AND MAINTENANCE

2.4.1 General

Systematic use of local disposal of stormwater is a relatively recent phenomenon. As a result, most installations in existence today have been in operation only a few years. The Environmental Protection Agency of the United States of America (EPA, 1983) reported considerable success in control of storm runoff and pollutants through the use of local disposal devices. Removal of pollutants over the long term was estimated to range between 45% and 65% without adverse contamination of groundwater.

Because of the relatively short experience, it is still too early to draw far-reaching conclusions regarding the installations' long-term reliability or life expectancy. The following sections discuss some of the construction, operation, and maintenance lessons gained from the experience so far.

2.4.2 Quality Control in Construction

Obviously, satisfactory performance of local disposal is very much dependent on the installation being built in accordance with the designer's plans, specifications, and/or instructions. Although this is obvious, it is very common to see improper installation of local disposal facilities.

Unlike the construction of large public works projects that have full-time professional inspection, local disposal facilities, because of their small size, will often not have full-time inspection. Even if they do, according to Shaver (1986), the inspectors are often not familiar with the technology and functional relationships of such facilities. We need to recognize that a typical construction worker cannot be expected to know the special needs of these type of installations.

When selecting local disposal as a functioning part of a community's stormwater management system, proper installation inspection and quality control procedures will also have to be provided by the community. Inspectors will have to be hired and trained. They will also need to be trained to call for expert help when unforeseen field conditions are encountered that may require design changes. The importance of attempting to properly deal, on a case-by-case basis, with the installation problems cannot be overemphasized. The successful performance of the entire stormwater system depends on a quality assurance program.

2.4.3 Safeguarding Local Disposal Facilities

Infiltration and percolation facilities are usually constructed at the same time other facilities and structures are constructed. The risk of damage to these facilities is the greatest at this time. To minimize these risks, the following are recommended:

1. Locate percolation basins away from roads or construction haul routes. Heavy vehicles traveling over these basins can cause the surrounding soil to flow into the pores of the rock media.
2. Minimize the sealing of infiltration and percolation surfaces by keeping traffic off those areas where they are to be built. Also, locate other activities that could seal soil surfaces (e.g., cement mixing, vehicle maintenance, etc.) away from these sites.
3. Since runoff from construction sites is heavily laden with fine, suspended solids which clog infiltration and percolation facilities, keep runoff out of these facilities until construction is complete.

2.4.4 Maturing of Infiltration Surfaces

Newly constructed infiltration surfaces may not have as rapid an infiltration rate as more mature surfaces. This is attributed to freshly compacted surfaces and immature vegetative cover. Figure 2.11 shows an example of water standing due to reduced infiltration capacity.

After infiltration surfaces undergo freeze and thaw and the vegetation's root system loosens the soils, infiltration rates tend to increase. Use of nursery grown turf in newly installed infiltration basins can achieve the highest possible infiltration rates in the shortest amount of time after construction.

Since infiltration rates in a new facility are likely to be less than anticipated in design, the downstream stormwater conveyance system may appear to be somewhat undersized. It is wise to account for this interim period and to either slightly oversize the local disposal facility or the downstream conveyance system.

Figure 2.11   Standing stormwater due to reduced infiltration capacity.

As the local disposal facilities begin to age, some of them will fail and will have to be repaired or replaced. When failures begin to occur, the downstream conveyance system will need to handle more runoff. If the conveyance system was not designed with this in mind, it will become inadequate to handle the increase in runoff.

After land development is completed, random erosion occurs at points of concentrated flow (e.g., at edge of pavement, at roof downspouts, etc.). The eroded soils are carried to the infiltration or percolation facilities and the infiltration and percolation rates are significantly reduced. Control of erosion is very important and can be accomplished through the installation of splash pads at downspouts, the use of rock or paved rundowns, and other measures. Erosion control can go a long way toward reducing the rate of deterioration of local disposal facilities.

2.4.5 Clogging

It is not possible entirely to prevent fine sediments from entering the infiltration facilities and eventually clogging the soil pores. How long it takes will depend on the porosity of the soil, the quality of the stormwater, and how often runoff occurs. It is possible to reduce the amount of sediment through mechanical separation, such as presettlement in a holding basin. The use of filtration in advance of percolation beds, mentioned earlier, is very effective in reducing clogging.

If clogging occurs shortly after installation of infiltration facilities, it can indicate excessive sediment loads. One should check for erosion in the tributary area and for any other sources of excessive sediment. Frequent clogging may also indicate the blockage of filters at the inlets to the infiltration or percolation beds. If this is the case, the problem can be corrected by simply cleaning out the filter media.

Of a more serious nature is the clogging that will occur over a long period of time. This is due to the accumulation of pollutants in the pores of the soil and in the percolation media. This could take several years, unless there are unusually heavy loads of sediment from the tributary basin. Heise (1977) reported on an infiltration system in Denmark installed in 1950. It was designed to intercept runoff from street surfaces. Although it functioned well at first, it lost all of its infiltration capacity in 20 years. Excavation revealed that the soils were impregnated with oily sediments. Instead of replacement, a storm sewer was installed.

2.4.6 Frost Problems

Frost in the soils can adversely influence the performance of a local disposal facility. It is possible, however, to design mitigating features into an installation, but their reliability has yet to be field verified. For example, it is possible to install insulation above the stone in a percolation pit, but the filter areas at the inlets can freeze and prevent water from entering the rock media itself.

Freezing can be an extreme problem in cold climates, especially for infiltration basins. When that is the case, sufficient storage volume on the surface of the basin needs to be provided to store winter snowmelt until spring thaws reopen the infiltration surfaces.

2.4.7 Slope Stability

Because local disposal artificially forces stormwater into the ground, the possibility of creating slope stability problems should always be considered. As the water infiltrates or percolates into soils, the intergranular friction in the soil can be reduced and previously stable slopes can become unstable. Slope failures in urban areas can be disastrous. A geotechnical expert should be consulted whenever local disposal is being considered, regardless of the slope of the terrain. Such expertise is especially important if slope stability may be affected.

2.4.8 Effects on Groundwater

The forced inflow of stormwater into the ground will affect the groundwater levels and water quality in the regions where it occurs. The impact on groundwater needs to be considered and accounted for in the design of buildings. As an example, buildings with basements may not be feasible if the groundwater levels are raised above basement floor elevations. This problem may be solved by the installation of underdrains. At any rate, local disposal will have an impact on groundwater.

Local disposal may also have an impact on the quality of groundwater and may be of particular concern where groundwater is used as water supply. Studies reported to date by the EPA (1983) and others suggest that groundwater recharged by nonindustrial stormwater may not have serious water quality problems. Water quality samples taken at several sites revealed that the groundwater underneath infiltration and percolation facilities meets all of the EPA's primary drinking water standards. These findings are still considered site-specific and, as a result, are inconclusive for all conditions.

Recent findings of groundwater contamination by organic toxicants, however, leave room for concern. As our society continues to use various solvents, herbicides, pesticides, and other potentially toxic, carcinogenic, or mutagenic chemicals, many of these chemicals will enter groundwater with infiltrated stormwater. Unfortunately, the potential of these chemicals being present in stormwater runoff is ever present. And although there is no evidence of widespread contamination, as long as these chemicals are in general use we need to consider their potential presence in the stormwater runoff.

2.4.9 Care and Maintenance

As with any facility installed by humans, regular care and maintenance is needed to insure proper operation. Unfortunately, maintenance is often neglected. Often, local disposal facilities are constructed as a part of a new development and become the responsibility of the individual property owner. Maintenance or operational instructions are often not provided to the new owners.

As these facilities begin to fail, the property owners may conclude that the site was improperly graded in the first place. This "problem" may be resolved by improving surface drainage through regrading the site, or by installing drainage pipes or swales. Obviously, this defeats the local disposal concept and results in increased loading on the downstream system. It may be possible to avoid such a scenario if local authorities mandate, by ordinance, that the new owners be notified at time of sale that they are now owners of a local stormwater disposal facility. Such formal notices should also contain the operation and maintenance instructions.

Infiltration facilities, to a large extent, depend on a healthy growth of vegetation to keep the infiltration surfaces porous. Maintenance of the infiltration areas needs to include the maintenance of the vegetation growing on the infiltration zone. As the facility ages and the surface soils become clogged, the top soil layers may have to be removed, replaced, and revegetated to restore their infiltration capacity.

Maintenance of percolation facilities needs to concentrate on keeping the inlet filters from being plugged. The filter fabric and/or sand layers need to be checked frequently and cleaned when found to be excessively clogged. After a number of years it may be necessary to replace the rock media and the adjacent soils because the very fine, unfilterable particles fill the pores. Regular maintenance of the inlet filters can significantly increase the time before the percolation bed has to be replaced.

REFERENCES

EPA, *Results of the Nationwide Urban Runoff Program—Final Report,* U.S. Environmental Protection Agency, NTIS Access No. PB84-185552, Washington, D.C., 1983.

HEISE, P., "Infiltration Systems," *Seminar in Surface Water Technology,* Fagernes, 21–23, March 1977. (In Danish)

SHAVER, H. E., "Infiltration as a Stormwater Management Component," *Urban Runoff Quality—Impact and Quality Enhancement Technology,* American Society of Civil Engineers, New York, 1986.

SWEDISH ASSOCIATION OF WATER AND SEWAGE WORKS, *Local Disposal of Storm Water—Design Manual,* Publication VAV P46, 1983. (In Swedish)

WILLIAMS, L. H., "Effectiveness of Stormwater Detention," *Proceedings of the Conference on Stormwater Detention Facilities,* American Society of Civil Engineers, New York, 1982.

3

Inlet Control Facilities

3.1 INTRODUCTION

Until the 1960s, a typical approach to stormwater drainage design was to convey the water away from where it fell as rapidly as possible. As a result, stormwater runoff was accelerated, and the flow peaks increased substantially as land urbanized. The concept of controlling runoff at its source became popular in the 1960s. The idea was to temporarily detain stormwater runoff at its source and to release it at reduced rates into the downstream conveyance system. The peak runoff rates from individual sites can be reduced by:

- increasing the travel time of runoff to the inlet, and
- detaining runoff at its source.

3.2 INCREASING THE TRAVEL TIME OF RUNOFF

The travel time of stormwater runoff can be increased by reducing the surface flow velocity and by lengthening the distance runoff has to travel. The velocity of flow on any surface can be approximated using Manning's Equation:

$$V = \frac{1.49}{n} R^{2/3} S^{1/2} \qquad (3.1)$$

in which V = velocity of flow, in feet per second,

$\quad\quad\quad n$ = Manning's roughness coefficient,

$\quad\quad\quad R$ = depth of flow, in feet, and

$\quad\quad\quad S$ = slope of the ground, in feet per foot.

The practical approach tells us that only two variables in this equation can be varied during design, namely, Manning's roughness coefficient n and the slope of the ground S. As an example, the slope can be reduced during site grading of new development. Reduction of slope will also increase depression storage of the surfaces. Surface depression storage retains some of the rainfall and keeps it from running off. The retained water forms puddles which eventually evaporate or infiltrate into the ground.

On the other hand, flat slopes can result in poor surface drainage, which can become a nuisance. Also, poor drainage of paved areas can reduce pavement life and increase its maintenance frequency. Such potential problems need to be evaluated before selecting what cross-slopes to use for paved areas.

It is much easier to increase the roughness coefficient of grassy areas than it is for paved surfaces. The slope of grassy surfaces can be made to undulate and thus slow down runoff velocities. On the other hand, paved areas, such as parking lots, can be interspersed with grassy areas to reduce the average flow velocities along the surface of the site.

These are simple concepts that are often not used because local drainage or paving criteria require curb and gutter at all pavement edges. Where permissible, these concepts offer possibilities for reducing surface runoff from developed sites; however, it is very difficult to actually quantify by how much. Instead of using curb and gutter, the edges of the paved areas can be delimited with precast wheel stops that permit drainage to flow onto the grassy surfaces as sheet flow.

Often, there are opportunities in site design to increase flow paths to storm sewer inlets. For example, sites can be graded to have slow surface swales that, more or less, parallel the contours and intercept runoff. Whenever this is possible, the runoff travel time is increased, runoff peaks are reduced, some of the runoff is infiltrated into the ground, and slope erosion problems are reduced. Creativity, within sound technological constraints, can help achieve good drainage, aesthetic enhancement, erosion control, and savings in the cost of construction, operation, and maintenance.

3.3 SURFACE STORAGE

More direct inlet control temporarily detains runoff at, or near, the source of runoff. Such detention can occur on flat roofs, parking lots, storage yards, and other surfaces specifically prepared for this purpose. Many of these techniques were originally introduced in the *Urban Storm Drainage Criteria Manual* originally published by the Denver Regional Council of Governments (1969). Since then, these concepts have proliferated worldwide, and examples of their use can be found in many of the industrialized countries in the world.

Source control has the potential, if properly implemented and maintained, of being an effective stormwater management technique (Urbonas & Glidden, 1983). However, one must be aware of its limitations and pitfalls, which are discussed in the following sections.

3.4 ROOFTOP STORAGE

Stormwater can be detained on a flat roof by installing flow restrictors on roof drains. Flat roofs are designed to hold a substantial live load and are sealed against leakage. Nevertheless, when a roof is used for detention, the structural design needs to account for the increased loading in accordance with the recommendations of the Uniform Building Code.

The Uniform Building Code also recommends minimum standards for the design of flat roofs, which include minimum roof slopes and maximum ponding depths. The designer should check the latest codes and standards before finalizing his or her plans. If all the UBC requirements are fully adhered to as they exist at the time this book was written, very little effective storage volume is available on rooftops. However, it is a concept that is often practiced and may be worth consideration.

A typical design for a flow restrictor that is used at a roof drain is illustrated in Figure 3.1. As can be seen, the outlet has a strainer that is surrounded by a flow restricting ring. The degree of flow control is determined by the size and number of holes in the ring. When the water depth reaches the top of the ring, it then spills freely into the roof drain with virtually no further restriction. Water ponding depth is thereby controlled to a permissible depth while providing a controlled release rate for a measured storage volume.

Rooftop detention is not without its problems. The most common is lack of proper inspection and maintenance. The flow control ring can clog with debris, such as leaves, and cause the water to pond for prolonged periods. Building owners have been known to remove these flow restrictors to eliminate the nuisance of ponding water on the roof, often not realizing that the control ring is an integral part of the community's drainage system. This happens frequently after a roof develops a leak.

A follow-up inspection and enforcement program, to insure that all roof restrictors are working, is generally not practical for a municipality. Thus, roof detention controls have a tendency to disappear with time, and their value as a stormwater management tool needs to be questioned. A routine municipal inspection and enforcement program is one way to minimize the loss of roof detention. However, commitment for such an inspection program is often not possible from the elected officials. As a result, rooftop detention cannot be expected to be effective with time in reducing flows in the downstream conveyance system.

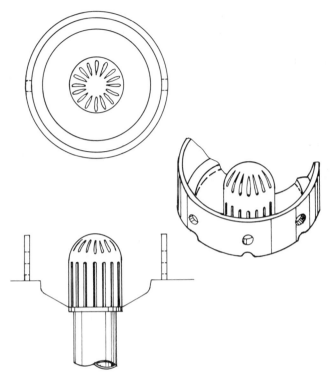

Figure 3.1   Roof detention drain control ring. (After Poertner, 1974.)

3.5 DETENTION ON PAVED SURFACES

Parking lots, paved storage yards, and other paved surfaces can be, and are, often used for stormwater detention. The advantage parking lots and other paved surfaces have over rooftops is that parking lots provide a much larger storage surface which can also pond to a greater depth. The use of parking lots as detention facilities has become very popular in some parts of the United States. The reason for this is that very little additional land needs to be dedicated exclusively for detention in a commercial development. As a result, this provides an economic incentive for the land developer to lobby for the acceptance and the continued use of parking lots as detention basins.

Unfortunately, as often happens with new concepts that catch on quickly, economic or political expediency can get in the way of sound engineering practice. Often, the decision makers embrace the idea of parking lot detention as a panacea for all drainage problems (Urbonas, 1985). As with rooftop detention, the use of parking lots for detention needs to be backed up with the staff resources to insure their continued existence and proper maintenance. Merely requiring their installation at the time of development is not enough.

Parking lot detention shares the same surface area with parked vehicles. If the detention is designed without regard for the primary use of the parking lot in mind, considerable inconvenience and damage to parked vehicles can occur when it rains. First and foremost, for the parking lot detention to be acceptable to its owners, it is necessary to insure that the lot does not pond water frequently. Also, when the lot detains stormwater, it should be inundated for only a short period of time. Thus, it is important for the designer to recognize the limitations in ponding depths and the frequency of ponding. Failure to do so can lead to owners taking action to eliminate this nuisance after experiencing flooding on their property.

Here are several rules for designing relatively successful parking lot detention. The same rules are also appropriate for detention on other types of paved surfaces.

1. Keep the frequency of ponding on the lot to a minimum. Ponding at full depth no more frequently than once every 5 to 10 years will keep the nuisance factor very low. This may require that detention of the more frequent smaller storms occur off the parking lot surface.

2. Keep the *maximum* depth of ponding during a major storm (i.e., 100-year storm) to less than 8 inches.

3. Locate the deepest ponding zones at remote and least used portions of the parking lot. Obviously, site layout may dictate otherwise; if so, ask the owner to determine if alternate site layout is possible.

4. Drain the parking lot ponding quickly, preferably in less than 30 minutes.

5. Keep the flow restrictors out of easy reach to reduce vandalism and to discourage owners from removing them. Use a buried pipe as the primary flow control device instead of an orifice plate, which can be easily removed.

It is clear from the preceding rules that there may exist a conflict between the stormwater management needs and these rules. For example, if the local requirements are to control frequently occurring storms (i.e., 2- or 5-year events), parking lot detention will result in frequent ponding. On the other hand, if the requirements call for the control of very large storms (i.e., 10- or 100-year recurrence frequency), then the frequency of ponding will be low.

Use of parking lots in conjunction with ponds adjacent to the parking lot or with underground storage should not be overlooked. The adjacent or underground detention facility can store the more frequent events permitting the runoff from the larger storms to back up onto the parking lot surface. As an example, this concept was incorporated into the Arapahoe County Storm Drainage and Technical Criteria (1985) in Colorado. These same criteria also incorporate all of the aforementioned rules for parking lot detention and provide for multifrequency and multistage control.

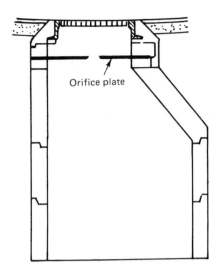

Orifice plate

Figure 3.2   Use of control orifice plate
inside an inlet. (After Poertner, 1974.)

There are many ways one can design a flow restrictor to control the
rate of drainage from a parking lot detention. One concept mentioned in the
American Public Works Association Special Report 43 by Poertner (1974) is
illustrated in Figure 3.2. This simple concept uses an orifice installed below the
grate of an area inlet. There are also specialty devices, such as the hydrobrake
and the flow control valve, that also can be used. These specialty devices will
be described in more detail in Chapter 12.

Whenever designing detention on paved surfaces that are used for other,
probably more noticeable purposes (e.g., parking of cars, storage of building
materials, storage of trucks or buses, etc.), the designer needs to provide
surface overflows for larger storms. These have to be built into the system to
act as a precaution against the possibility that the primary outlet may become
clogged, or a larger-than-the-design storm may occur. Failure to incorporate
an emergency overflow can cause damage and result in liability that can
otherwise be avoided. Even these precautions may not prevent damage during
unusually large rainstorms, but during such events, damage is likely even if
detention is not a part of the system.

REFERENCES

APWA, *Urban Stormwater Management,* Special Report No. 49, American Public
Works Association, Chicago, Ill., 1981.

ARAPAHOE COUNTY, *Storm Drainage Technical Criteria,* Arapahoe County, Colo.,
1985.

DRCOG, *Urban Storm Drainage Criteria Manual,* Vols. 1 & 2, Denver Regional
Council of Governments (currently being published by Urban Drainage and Flood
Control District), Denver, Colo., 1961.

POERTNER, H. G., *Practices in Detention of Urban Storm Water Runoff,* American Public Works Association, Special Report No. 43, 1974.

URBONAS, B. R., "Stormwater Detention, Idealism of 1960's Matures," *Proceedings of the Seminar on Flood Plain Hydrology,* Dept. of Civil Engineering, New Jersey Institute of Technology, Piscataway, N.J., 1985.

URBONAS, B. R., AND GLIDDEN, M. W., "Potential Effects of Detention Policies," *Proceedings of the Second Southwest Region Symposium on Urban Stormwater Management,* Texas A&M University, November 1983.

4

Open Ponds

4.1 INTRODUCTION

It is generally acknowledged by the stormwater management profession that urbanization increases runoff from rainstorms and snow melt. For example, studies in Denver by Urbonas show that during a routine summer afternoon rainshower, an acre of pavement will produce the same amount of runoff as would occur from several square miles of native rangeland. During an 1-inch thunderstorm, one paved acre may yield the same amount of runoff as 40 to 100 acres of rangeland. Clearly, urbanization radically increases storm runoff for the majority of storm events. This has a very significant effect on downstream residents and drainage systems. Detention can help mitigate, but not totally eliminate, these effects.

Open ponds are probably the most common type of detention used in stormwater management. They have the advantage of being dedicated totally to the task of stormwater management, which is not the case for rooftops or parking lots. They are also visible. This attracts attention for maintenance. Although there are examples of using open detention for control of combined sewer overflows, public health and aesthetic considerations generally limit the use of open detention ponds to the control of separate stormwater runoff.

4.2 PLANNING FOR DETENTION PONDS

When planning for a detention pond, it is not sufficient to address only hydrology and hydraulics. According to American Society of Civil Engineers (ASCE) (1985), successful detention facilities also have strong recreational or other community uses. The detention aspect is often considered secondary by the residents in the area. For these reasons, planning for detention needs also to consider the social, environmental, and recreational needs of each community.

When open ponds are planned, an attempt should be made to combine the detention use with other community uses. The following sections discuss such co-uses, along with the design considerations for each.

4.2.1 Effects on the Landscape: Aesthetics

As an integral part of the community it serves, a detention pond needs to blend into the landscape and into the community. Too often, detention ponds are installed merely as a hole in the ground (sometimes referred to as an HIG) without any redeeming landscape features.

Simple yet inexpensive measures, such as gentle side slopes, planting of trees and shrubs, and other landscaping features can transform an HIG into an attractive amenity for the neighborhood. The services of a competent land use planner and a landscape architect early in a development's planning process can help achieve more useful and attractive results.

4.2.2 Pond Environment

Ponds that have a permanent pool of water offer many attractive environmental possibilities. As urbanization occurs, there is a loss of wildlife and bird life habitat. Such habitat is replaced by manicured lawns, shrubs, and trees that offer habitat for select small birds and animals such as squirrels. It is possible to create a natural micro-environment around ponds that attract a greater variety of animal and bird life. These "natural" environmental pockets are considered by many city dwellers to be a treasure in an otherwise densely urbanized community.

Proper planning for riperian vegetation that fits the urban setting needs to be considered when designing the pond's landscape. As an example, water lilies and cattails can contribute to pond's aesthetic appeal. On the other hand, the uncontrolled spread of wild vegetation can be a nuisance to the surrounding neighborhood and become an operational and maintenance problem. Some of these problems are discussed later.

4.2.3 Recreational Opportunities

Detention basins and ponds, with or without permanent pools of water, offer many recreational opportunities in an urban setting. In Figure 4.1 we see two examples of detention ponds in Sweden. Both were planned and designed in cooperation with park officials of each community. These two examples illustrate how stormwater detention can be incorporated nicely into the landscape and provide recreational opportunities for local residents between storms. Since the periods between storms generally far exceed the periods of rainfall, these facilities are available for recreational uses most of the time.

Two other examples of active and passive recreational opportunities are illustrated in Figures 4.2 and 4.3. The first shows a detention basin in the central business district of Denver. It is used by people during the day for

Figure 4.1   Examples of ponds in Sweden at two parks.

Figure 4.2   Skyline Park in downtown Denver.

Figure 4.3 Englewood High School athletic field. *(Photo courtesy of McLaughlin Water Engineers, Ltd.)*

strolling, eating lunch, or just relaxing in the sun. The attractive landscaping fits the surrounding architecture and the inner city setting. Most people that use it do not realize that this three-block long basin is designed to actually reduce the storm flow rates to the available capacity of the downstream storm sewers. The second example is also in the Denver area and shows where a detention basin is used as a high school tournament athletic field. It is an off-line pond that controls floods downstream and stores water when the adjacent channel carries floods that exceed the 10-year storm.

Other examples, not illustrated here, include ponds with large permanent water surfaces that are used for skating in winter and for boating in summer. The possibilities for multiple use are limited only by the imagination and the community's desires to combine stormwater detention ponds with uses other than just storing water temporarily when it rains.

4.2.4 Removal of Pollutants

Detention basins and ponds, because they back up water, will cause suspended solids to settle. Since many of the pollutants are attached to suspended solids, ponds will remove some of them. How much is removed will depend on pond volume, inlet and outlet configuration, pond depth and shape, the time the stormwater resides in the pond, and whether or not the detention facility has a permanent pool of water.

Considerable information has been developed in recent years on the design of detention facilities for water quality enhancement. Nevertheless, we still have a lot to learn before we can design them for this purpose with complete confidence that we know what we are doing. Since this is a topic of interest to many stormwater managers, Part 4 of this book is dedicated to the design of detention for water quality enhancement.

4.2.5 Detention in Natural Lakes

When the downstream recipient of urban storm runoff is a natural lake, a water supply reservoir, or a recreational reservoir, each of them can provide peak flow attenuation. The flow routing advantages of these water bodies, however, can extract a price in the form of water quality deterioration and adverse impacts on their natural or designated uses. Potential causes of water quality degradation and resultant concerns may include the following:

- Nutrient enrichment resulting in accelerated eutrophication. Excessive algae levels can deplete oxygen and cause fish kills. According to Wallen, Jr. (1984), these factors also have been linked to impairment of recreational uses.
- Deposits of sediments containing heavy metals and attached petroleum product will occur in the bottom.
- If salt is used to control street icing, increases in lake salinity can occur. So far, there is little documentation of this actually becoming a serious problem.
- If acid rain is of concern, the increased surface runoff from urbanization may increase the acidity of the receiving water body.

4.2.6 Safety

When we discuss safety of detention ponds or basins we are, in fact, discussing a wide range of possibilities. This topic encompasses the structural integrity of the confining embankment and of the outlet works and the safety of the people using the facility for recreation. The latter includes the need to protect people when the pond is storing runoff (i.e., operating) and during the periods between storms.

The protection of embankments against catastrophic failure due to water overtopping them are discussed in more detail later in this section. At this time, we focus on the day-to-day safety considerations for the public in the vicinity of a pond.

The designers, owners, and approving agencies need to recognize that detention ponds are a part of the total urban community's infrastructure. As a result, they need to be designed with public safety in mind. It is inevitable that people will have access to these facilities, especially if the detention facility is in a park.

We have seen public works officials overreact and completely fence detention facilities. Fortunately, there are many more examples, such as reported by Edwards (1982), Stubchear (1982), and others where detention ponds are multiple-use facilities. Although safety should remain a concern, a properly designed detention pond should be no more hazardous than an urban lake, a playground, a hiking trail, or any other recreational or park facility in a city. It definitely is a safer facility than a city street, yet no one fences off streets.

Probably the most reliable safety feature to protect the public against accidentally falling into a pond is to install gentle side slopes for all the banks and the embankment. Banks should be sloped no steeper than 4 units of measure horizontally for each unit vertically (i.e., 4:1) Slopes steeper than 3:1 should be avoided entirely unless the facility is fenced. Gentle side slopes for the first 10 to 20 feet under water facilitate easy escape if a person accidentally enters the pond.

ASCE (1985) recommends that the designer pay particular attention to safety features around the outlet works. It recommends that outlet risers be located away from shore to discourage the public from access. Also, installation of trash racks with low entry velocities will contribute to safety during operation. Fencing, on the other hand, is less effective than it first appears, since the public, especially children, have been seen to bypass fences to access outlet areas.

Wherever possible, one should minimize the visibility of the outlet, thereby reducing its attraction potential. In some cases, installing signs warning of the hazard may be appropriate; however, signs also appear to have a limited record of success in keeping people away from the more hazardous areas. Clearly, the exact methodology will vary with its urban setting.

4.2.7 Layout of Detention Ponds

When planning a detention basin, try to lay it out so that it fits the surrounding landscape and the community. Unfortunately, there are too many examples of treating detention facilities as though they were a nuisance. As a result, they are sometimes located and designed without any regard for the community's needs or how the pond or basin may enhance the aesthetics of the surroundings. A detention facility can be designed to be an attractive and aesthetic amenity in an urban neighborhood. A side benefit of incorporating an attractive detention facility into the community's landscape is that it is more likely to receive proper maintenance.

Detention ponds should be laid out to insure that the flow entering the pond is evenly distributed across the pond so that stagnant zones do not develop in the pond. Zones of stagnation tend to become overgrown with vegetation and can increase mosquito breeding. Oblong shapes, with inlet and outlet at opposite ends, appear to be best suited for this purpose. If this shape is not possible, an elongated triangular shape with the inflow at its apex provides a reasonable alternative.

4.3 TECHNICAL CONFIGURATION

Technically, there are two types of detention facilities: those that store water only when stormwater is being detained and those that have a permanent pool of water. The latter store stormwater above the water surface of the permanent pond. We discuss technical considerations for both types of detention facilities.

4.3.1 Inflow Structure

Erosion and sediment deposition problems can develop at the inflow to the detention basin. Although it is possible to design inflow structures to minimize erosion, deposition cannot be prevented. To minimize maintenance costs, however, it is a good idea to localize much of the deposition where it can be easily removed.

To address the concerns we have just discussed, inflow structure needs to accomplish the following and have the following features:

- Dissipate flow energy at the inflow.
- Drop the inflow elevation when it enters the pond above the pond's water surface.
- Accelerate the diffusion of the inflow plume.
- Provide protection against erosion.
- Provide maintenance access for the repairs to the inlet and for the removal of sediments.
- Incorporate safety features to protect the public (i.e., gentle slopes, fencing or railing at vertical faces of the structure).
- Be unobtrusive to the public eye by blending the inlet into the surrounding terrain.

Inflow energy can be dissipated with the installation of standard energy dissipating structures. Although meant for much larger installations, some of the spillway structures developed by the U.S. Bureau of Reclamation (1964, 1974) provide good examples of design. Of course, the designer may have to scale down some of these designs. These types of structures can be modified in appearance to fit the urban setting. A good example of an attractive baffle chute located in a park is illustrated in Figure 4.4.

Typically, baffle chutes, vertical spillways, and sloping chutes with energy dissipating basins work well as inlet structures when the inflow has to be dropped in elevation. The reader is encouraged to study the cited references and modify as needed the standard designs to fit each site. One word of caution, however: It takes considerable mass to dissipate flow energy normally found at even small installations. Do not skimp on these structures. Unless properly designed and constructed, they can fall apart very quickly.

A settling basin or a settling zone should be provided near the inlet. The heaviest sediments will settle out at this location, and their removal thereby will be facilitated. A simple design for a settling basin is to depress the bottom near the inflow structure. Stabilize the bottom of this depressed area with soil cement, prefabricated slabs, or concrete paving; however, this is not abso-

Figure 4.4 Baffle chute energy dissipator in Denver.

lutely necessary. A stabilized bottom provides the maintenance crews with fixed boundaries within which sediment removal is to take place.

4.3.2 Configuration of Pond Bottom

The configuration of the pond bottom will depend, to a large degree, on whether the pond will have a permanent pool of water. Bottoms with permanent pools are, in some respects, easier to design. This is especially true if the low flows are conveyed in a pipe to the permanent pool. The areas above the permanent pool can then be graded relatively flat, kept clean with less effort, and can be made available for other uses (e.g., play fields, picnic tables, passive recreation, etc.) for longer periods of time.

The permanent pool can also act as the settling basin. The EPA (1983) reported that a permanent pool will cleanse stormwater of pollutants much more effectively than a detention basin without a permanent pool of water. Thus, if water quality enhancement is a goal, this configuration offers proven advantages.

Vehicular access needs to be provided to the entire perimeter of the pool to facilitate maintenance. If possible, provide gates or valves to totally drain the permanent pool. When drained, the bottom is much easier to clean and to excavate than when it is under water. This translates into maintenance cost savings.

A detention basin that stores water only during storms should have a trickle flow ditch between the inlet and the outlet. Two examples are illustrated in Figure 4.5 in which the low flow ditches are sized to carry the frequently occurring runoff and trickle flows. When large rainstorms occur, the capacity of these ditches is exceeded and the water floods the adjacent pond bottom.

The most successful installations of low flow and trickle flow channels have a concrete bottom. Concrete lining facilitates self-cleansing of the ditch

Figure 4.5   Two examples of trickle flow channels.

and its maintenance. Drainage of the pond bottom between storms has to occur efficiently if it is going to be used for recreation. To achieve this, the bottom has to be cross-sloped at no less than 2% toward the trickle channel(s). Where high groundwater is present, the bottom may not be suitable for passive or active recreation unless subdrains are installed.

Opportunities for multiple use may be plentiful, but it is up to the designer to insure that multiple uses will indeed be possible. Poorly configured and poorly drained detention pond bottoms can foreclose recreational uses, even when such uses are desired or intended at the detention pond.

4.3.3 Slope Protection

Side slopes of the pond will tend to erode whenever the detained water surface fluctuates frequently or when there is wave action. Good vegetation will help to protect the side slopes against erosion; however, in areas of high velocities and wave attack, structural measures are needed to supplement vegetation. Figure 4.6 shows an example of rock slope erosion protection in an overflow zone. Other techniques include concrete lining, burried riprap, and soil cement.

An example of bank protection at the water level is illustrated in Figure 4.7. This type of erosion protection has to be provided when the ground slope under the water is steeper than the normal beaching slope. The use of rock or

Figure 4.6 Slope protection near an outlet.

Figure 4.7 Slope protection at the normal water level.

gabion has been generally successful for this purpose, while the use of hard linings, such as concrete, has had only a marginal success record.

Structural concrete, with adequate bedding and subdrainage, should also perform well as slope protection. In designing concrete slope protection, the forces of wave action and those of freeze and thaw can displace concrete slabs. The U.S. Bureau of Reclamation (1978) provides detailed design suggestions for the design of slope protection that should work well for larger urban detention reservoirs. This and other publications by the Bureau provide excellent guidance for a designer of detention ponds and basins.

4.3.4 Outlet Structure

The configuration of a pond outlet determines the type of pond (i.e., wet or dry), the storage volume, and the control the pond provides the storm runoff. For some time now, many detention ponds were being designed to control runoff from a single recurrence frequency of rainstorm. As a result, some ponds were sized to control, for instance, only the 100-year storm, the 10-year storm, or some other design storm. Studies by Kamadulski and McCuen (1979) and Urbonas and Glidden (1983) concluded that the control of a single frequency of runoff will not effectively control storm runoff of a different runoff probability.

The recommended method is to provide outlets to control the flows of at least two recurrence frequencies of runoff. Brulo et. al. (1984), Kamelduski and McCuen (1979), and Urbonas and Glidden (1983) have described the advantages of two-staged outlets in controlling multiple flood frequencies. Control of two widely different recurrence frequencies of storms can control runoff of other recurrence frequencies as well. It is possible to design outlets that can provide more levels of control, but each level of control adds complexity and cost. From a practical perspective, design for two levels of control, such as the 2-year and the 10-year floods, should be sufficient in most settings. If major floods are of great concern, the outlet can be designed to control the 100-year flood in addition to the other two events.

Although the goals may vary, the most common practice is to control the release rates so that the flows after land development are no larger than before land development. This is frequently specified by local ordinance or criteria which also specify the design storm or storms that need to be so controlled. The discharge rate specified by such criteria is the discharge that should occur when the storage volume occupies the volume allocated for the storm of specified recurrence frequency. If more than one level of control is desired, there is a unique design discharge that may be released for each level of storage from multiple outlets.

In designing outlets, consider the following:

- Design the outlet control orifice in a way that makes unauthorized enlargements impractical. Use of a pipe section that limits the flow rate instead of an orifice plate is one method for achieving this.

- Design the outlet for maximum safety to the public.
- Wherever possible, design for the control of two or three levels of flow (e.g., 2- and 10-year; 2- and 100-year; 10- and 100-year; 2-, 10- and 100-year; etc.).
- Provide maintenance access to the outlet.
- If possible, use no moving parts or pumps in an outlet.
- Use massive components to reduce damage from vandalism.
- Provide erosion protection at the inlet and outlet ends of the outlet pipe.
- Provide coarse gravel packing to screen out debris whenever a perforated outlet riser is used in a dry detention basin.
- Provide a skimmer type shield around a perforated riser in a wet pond to skim off floating debris.
- Always design with maintenance and aesthetics in mind.

Figures 4.8 and 4.9 contain two examples of outlets for dry detention basins. These outlets will permit the entire pond volume to drain between storms. Both provide a small drop between the pond bottom and the lowest flow control orifice to improve the drainage of the bottom.

An example of an outlet for a pond with a permanent water pond is seen in Figure 4.10. After the storm is over, the water surface eventually drops to the permanent pool level. This level is controlled by positioning the openings at the desired permanent water level. Regardless of the configuration, install a gate or a valve near the bottom of the permanent pool. This will greatly facilitate sediment and debris removal from the bottom and its regular maintenance.

4.3.5 Trash Racks

As a practical matter, outlets should be provided with a trash rack to prevent plugging with debris and to provide for safety to the public. An example of a trash rack for a dry pond is illustrated in Figure 4.11. This example shows a rack that is sloped to facilitate cleaning and to reduce entrance flow velocities. Experience suggests that sloping of the rack at 30% to 50% above the horizontal plane seems to work quite well and permits "raking off" of floatables during operation.

For very small control outlets (i.e., less than 6 inches in diameter) the trash rack needs to have a relatively large opening. Kropp (1982) suggested that small orifice outlets be protected with a rack having a net opening of no less than 20 times the opening of the orifice. He did caution, however, to keep the spacings between openings narrower than the orifice itself. If the openings are spaced too far apart, the trash carried by stormwater can pass through the rack and plug the orifice.

For larger outlets, namely 24-inch diameter or larger, good operational results can be obtained when the rack's net opening is no less than four times

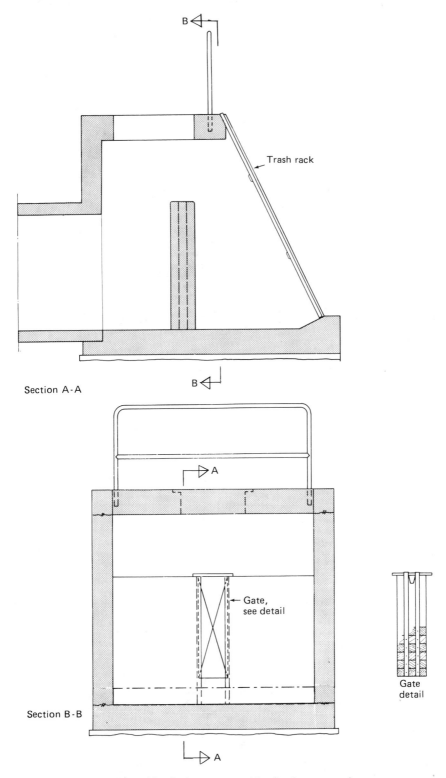

Section A-A

Trash rack

Section B-B

Gate,
see detail

Gate
detail

Figure 4.8 Outlet structure with a fixed gate control.

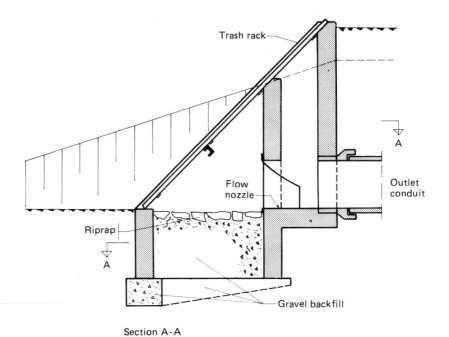

Trash rack

Flow nozzle

Riprap

Outlet conduit

Gravel backfill

Section A-A

Control nozzle of stainless steel

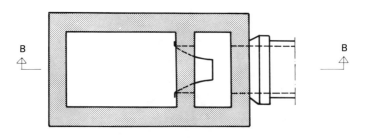

Section B-B

Figure 4.9 Outlet structure with a fixed orifice control.

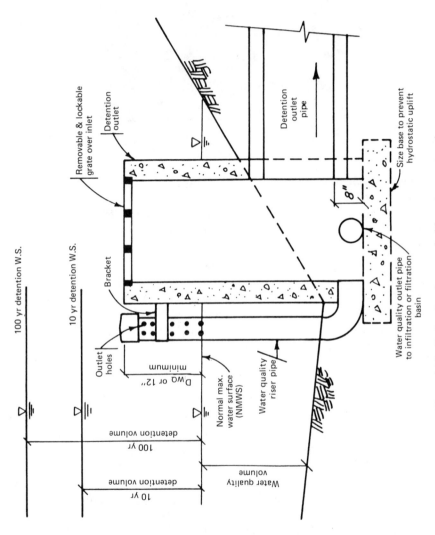

Figure 4.10   Outlet riser for a pond with a permanent pool.

45

Figure 4.11   Example of a trash rack for a pond outlet.

the opening of the outlet. For example, a 24-inch diameter outlet can be reasonably protected by a sloping trash rack having a net opening of 12.5 square
feet. Although this may sound large, a rack approximately 4 feet on each side
will provide more than adequate protection to the public and keep the outlet
from plugging.

Consider the following when designing trash racks for dry ponds:

- For outlets with 6-inch diameter openings or less, provide a net area of
  trash rack that is no less than 20 to 40 times the area of outlet.
- The spacing between the openings of the trash rack should be smaller
  than the smallest dimension of the outlet.
- For outlets 24 inches or more, provide a net area of the rack that is no
  less than four times the area of the outlet.
- For larger outlets, the spacing between the openings of the rack should
  be 6 inches or less to prevent a child from passing through the rack.
- Slope all racks at 30% to 50% above the horizontal to facilitate cleaning.
  A flat angle also permits the floating debris to "ride up" the rack as the
  pond level rises, thereby freeing up openings underneath.
- For larger outlets, a 4- to 6-inch horizontal opening at the bottom of the
  rack will permit the smaller debris to flush through the outlet.

4.3.6 Spillway vs. Embankment Overtopping

In the 1985 report to the American Society of Civil Engineers by a task
committee set up to investigate the state of the art of detention pond outlets,
safety of detention outlets was particularly stressed. In addition to public
safety, the committee expressed a strong concern for the protection of the

detention pond dam against failure due to overtopping and due to failure at the outlet.

A detention pond behind an embankment is nothing more than a reservoir of water behind a dam. The scale may be much smaller, but the analogy holds. As a result, a breached embankment may be the cause of much more damage than would occur from flooding without the detention pond. Fortunately, many detention ponds are so small that embankment safety does not have to be an issue of great concern.

Always consider the probability of embankment failure and the circumstances under which such failure is possible when designing a detention facility. Once the overtopping flood has been identified, such as the 500-year flood, estimate the consequences of failure in light of the flooding that would occur from such a storm if there was no dam. Will the failure be rapid or will the embankment breach slowly? Will the predicted mode of failure increase the flooding damages, and to what degree? Your analysis may reveal that despite failure, flood damages would be essentially the same as those that would occur if the detention pond was not present. With this type of analysis, the designer is in a strong position to make rational and cost-effective recommendations.

An emergency spillway is always a good idea. It can be designed to protect the embankment from overtopping whenever storms larger than design storms occur or when outlets plug with debris. Clearly, a spillway designed to pass the probable maximum flood is desirable but may not be practical or possible to provide for a small urban pond. As a result, spillways for smaller ponds are often sized for less than the probable maximum flood. In those cases, the embankment can be constructed to resist sudden failure during rare and extremely large floods. One example of an attractive emergency spillway for a small detention dam is shown in Figure 4.12.

An embankment overtopping analysis (described in the next section) along with incremental flooding damage analysis for floods up to the maximum probable flood can serve as a basis in setting the level of protection at each site. Under no circumstances, however, should the emergency overflow from a pond be designed to have a direct path to any buildings or other structures used for human occupancy, commerce, or industry.

4.3.7 Embankment Loss Analysis

Urban watersheds are typically small, and the amount of total time an embankment is actually overtopped is much shorter than for a large watershed. A detailed embankment loss analysis can quantify the breaching potential of an embankment for a variety of large storm events. A procedure based on the results of highway embankment overtopping studies by Chen and Anderson (1986) for the Federal Highway Administration could be used for this purpose.

Figure 4.12   An example of an emergency spillway.

Figure 4.13 shows four nomographs that can be used to estimate the soil loss rate of an overtopped bare soil embankment. Two of these provide the erosion rates for a 5-foot high embankment, while the other two give correction factors for embankments under other design conditions. The procedure for estimating the erosive loss of soil is as follows:

1. Determine the height of the embankment and its soil type (i.e., cohesive or noncohesive).
2. Determine the headwater depth, h, and tailwater, t, above the embankment crest.
3. Compute t/h.
4. Using h and t/h, determine the erosion rate, E_a, from Figure 4.13 for the embankment soil type.
5. Again using Figure 4.13, determine the adjustment factors for embankment height and duration of flow.
6. Compute the erosion rate: $E = K_1 K_2 E_a$.
7. Compute the soil loss for the period the embankment is estimated to be overtopped.

The foregoing procedure may or may not provide an accurate estimate for a particular site. There are many uncertainties in its application, and it is

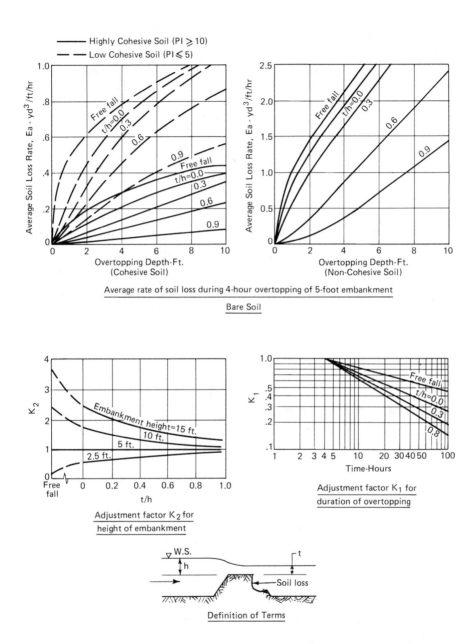

Figure 4.13 Soil erosion loss nomographs for overtopped bare soil highway embankments. *(After Chen and Anderson, 1986.)*

wise to provide safety factors in design. For example, when estimating the embankment loss, you may want to assume that overtopping occurs over only a portion of the crest length.

Here are some suggestions for how to mitigate possible adverse impacts of detention pond dam overtopping whenever the emergency spillway is of less capacity than the probable maximum flood:

- Construct the embankment using erosion resistant material. For example, concrete rubble or large rock embedded in compacted clay soils on the downstream side of the dam may be able to withstand considerable overtopping for brief periods of time without sudden and catastrophic failure.
- Use noneroding materials on the downstream portion of the embankment such as soil cement, rollcrete, interlocking concrete elements or concrete paving.
- Design the embankment with an extra wide crest or construct a paved road on top.
- Flatten the downstream face of the dam and revegetate with native turf-forming grasses.
- Enlarge the conveyance capacity downstream of the dam.
- Excavate the pond to eliminate or reduce the height of the embankment.

4.4 OPERATION AND MAINTENANCE

If you do not plan to maintain it, do not build it. That should be the golden rule in public works. As a result, all detention ponds should have maintainability built into them. This includes good maintenance access to all parts of the ponding area for mechanized maintenance equipment. Although this suggestion may seem trivial, one of the most frequently observed deficiencies is inadequate maintenance access. Can it be built, and can it be maintained? These two questions should always be asked as the detention facilities are being designed. If the answer is no to either question, modify the design.

In this section we discuss some of the operation and maintenance considerations for detention ponds. The following topics are discussed:

- general inspection and maintenance,
- algae and aquatic plants,
- sediments,
- floating debris and pollutants,
- general housekeeping, and
- erosion.

4.4.1 General Inspection and Maintenance

The responsibility for maintenance rests with the owner of the detention pond. As part of the design, provide an operation and maintenance manual to the owner. It is a good idea for the owner to have one prepared for already existing facilities. Such a manual does not need to be complicated. It merely needs to serve as a checklist of things to look for, and to do, on a regular basis. A more detailed, but more valuable, manual could contain copies of as-built drawings for key elements of the pond, a list of replacement parts, emergency procedures in case of very large floods, names and telephone numbers of key individuals, etc. Obviously, the level of detail will vary with the size and the complexity of the installation.

At the beginning of each flood season, and after each significant storm, inspect the detention pond for damage. Pay particular attention to the embankment (i.e., dam), the outlet works, the emergency spillway, and the inlet works. Look for signs of erosion, excessive deposition of sediments, buildup of trash or debris, or any other signs of damage. Address problems immediately. What may appear to be minor damage at the time can become the weak link if another flood occurs before repairs are made.

4.4.2 Algae and Aquatic Plants

Algae and other aquatic plants will only occur in "wet" ponds, namely in ponds with a permanent water pool. Since detention ponds often receive nutrient enriched water, the possibility of them having algae and/or other aquatic plants is indeed high.

Bottom vegetation, on the other hand, occurs mostly in the shallow areas of the pond. It is thus possible to control the extent of bottom vegetation through the shaping of the pond. If it is desirable to have vegetation near the shore for aesthetics, water quality, or aquatic habitat, the pond can be made shallow near the edges. At the same time, water depths exceeding 6 feet can be provided further from shore portions of the pond, thereby reducing the vegetation growth potential further from pond edges.

Although it is virtually impossible to prevent algae from growing in an urban detention pond, excessive bottom vegetation growth can be controlled through the design of the facility and by occasional harvesting. Although there are chemicals that can kill or control algae and other vegetation, their use can contaminate the receiving waters. For this reason, algae and vegetation control chemicals should be used with extreme caution.

Wherever the state wildlife or federal regulations permit consider the use of fish that eat algae and other aquatic vegetation. Their use has been occasionally successful in controlling excessive aquatic plant growth. Check with your state's game and fish personnel before deciding on their use. It may be wise to experiment first on a smaller installation to see if the fish improve the conditions in your part of the country.

Excessive algae can cause odor and aesthetic problems. Although total elimination of algae is virtually impossible, it is possible to mitigate their adverse impacts. Making the pond deeper can help in some cases, especially if the water in the pond is flushed through on a rather frequent basis. Also, the use of mechanical aerators can help mitigate the odors and reduce algae growth. If odors and excessive algae become a problem, draining the pond and cleaning out its bottom will remove some of the nutrient source entering the water column. Although not totally foolproof, this can reduce the problem until the bottom nutrients have again accumulated.

Dry detention basins should not have problems with algae growth. If, however, the basin does not drain fully, or the bottom is so flat that the bottom drains very slowly, marshy or wetland type vegetation can develop. It is not practical to mechanically mow the pond bottom when that happens, and very little routine daily maintenance can be done on the bottom of the basin. Occasionally, the entire bottom can be "mucked out" to remove excess sediment buildup, thereby cleaning the bottom every few years. If the stated goal is to reduce or to eliminate the marsh bottom conditions, then steepen the bottom slope when grading it or install subdrains to drain the bottom.

4.4.3 Sediments

Storage of runoff in basins will cause sediments to settle within them. The amount and rate of sediment deposition is a function of the source and quantity of sediments in the incoming water and their settling velocities. The rate of sediment buildup inside a basin or a pond will depend to some degree on how much the stormwater is slowed in the basin and how long the water stays in the basin.

Sediment deposition is particularly severe when land development activities are occurring in the tributary watershed. Soil erosion during the land development period, if not controlled, can quickly fill a detention pond. For this reason, it is strongly recommended that erosion control within the tributary watershed be practiced and that the detention ponds and basins be reexcavated on completion of the land development work. This practice will restore the pond's original design volume and insure that it will perform as originally designed.

The deposition of sediments in a detention basin does not occur uniformly over the bottom. The larger and heavier sediments drop out near the inlet to the pond. As the distance from the inlet increases, the bottom deposits consist of the smaller particle fraction of the sediment load. To facilitate maintenance and the removal of the deposited sediments, it is a good idea to incorporate a sediment trap near the inlet.

Removal of sediments is easiest when the pond bottom is dry. This can be scheduled during the dry season, when the pond can be drained completely. Allow sufficient time to dry the bottom soils so that mechanical excavating equipment can be used. Facilitate access for the mechanical maintenance equipment by installing access ramps that lead to the bottom of the pond. An example of such an installation is illustrated in Figure 4.14.

If it is not possible to drain or to pump the pond dry, underwater excavating techniques have to be employed. Examples of underwater excavation include hydraulic dredging or the use of a drag line or a "clam shell." Underwater excavation is much more expensive than dry excavation, a fact that needs to be recognized at the time of design. Consult with a geotechnical engineer to ascertain if the soils in the pond bottom will stabilize sufficiently, within a reasonable time after the pond is drained, to support excavation equipment. If the soils will not stabilize sufficiently, underwater excavation may be the only option.

4.4.4 Floatables

Detention pond and basin maintenance also includes the removal and disposal of floatables such as tree branches, lumber, leaves, styrofoam, litter, etc. carried to the pond by stormwater. In ponds that are designed to reduce peaks of larger storm hydrographs, the water moves rapidly through the pond and much of the floatable mass can be flushed through. In such cases, all that may be needed is an occasional debris collection and cleaning of outlet trash racks between storms.

On the other hand, floatables can be a problem for ponds (i.e., basins with a permanent pool) and for ponds and basins intended for water quality enhancement. Frequent trash pickup along the shore is needed to maintain a

Figure 4.14 Access ramp for maintenance equipment.

clean appearance. It is possible to install floating skimmers at a forebay into which the inlets empty stormwater runoff. This keeps the floatables from entering the main body of the pond, and the floatable trash is concentrated for easier removal and disposal.

4.4.5 General Housekeeping

To the public, the most important feature of any urban facility is a clean, well-maintained appearance. Because of this, it is wise to schedule frequent debris and trash removal and regular mowing of the grass around the pond and within the detention basin.

The frequency of such maintenance will be governed by the specific uses of detention facilities and whether they are located close to residential or commercial areas. One cannot offer standardized guidelines for scheduling such activities, since they vary between installations and between various communities. What is important, however, is to anticipate and to budget for routine maintenance and housekeeping when new detention facilities are being installed.

REFERENCES

ASCE, *Stormwater Detention Outlet Control Structures,* Final Report of the Task Committee on the Design of Outlet Structures, American Society of Civil Engineers, New York, 1985.

BRULO, A. T., KIBLER, D. F., AND MILLER, A. C., "Evaluation of Two Stage Outlet Hydraulics," *Proceedings of the ASCE Hydraulics Division Conference on Water for Resources Development,* New York, 1984.

CHEN, Y-H, AND ANDERSON, B. A., *Analysis of Data In Literature for Estimating Embankment Damage Due to Flood Overtopping,* Simons, Li & Associates, Inc. Report to U.S. Department of Transportation, Federal Highway Administration, DC-FHA-01, Washington, D.C., 1986.

EDWARDS, K. L., "Acceptance and/or Resistance to Detention Ponds," *Proceedings of the Conference on Stormwater Detention Facilities,* American Society of Civil Engineers, New York, 1982.

EPA, *Results of the Nationwide Urban Runoff Program,* NTIS Access Number: PB84-185552, Environmental Protection Agency, Washington, D.C., 1983.

KAMELDUSKI, D. E., AND MCCUEN, R. H., "Evaluation of Alternative Stormwater Detention Policies," *Journal of Water Resources Planning and Management Division,* Vol. 105, pp 171–86, American Society of Civil Engineers, New York, Sept. 1979.

KROPP, R. H., "Water Quality Enhancement Design Techniques," *Proceedings of the Conference on Stormwater Detention Facilities,* American Society of Civil Engineers, New York, 1982.

POERTNER, H. G, *Practices in Detention of Urban Stormwater Runoff,* American Public Works Association, Special Report Number 43, 1974.

SCHROEDER, G., *Agricultural Water Engineering,* 3rd ed., Springer-Verlag, Berlin, 1958. (In German)

STUBCHEAR, JAMES M., "Stormwater Basins In Santa Barbara County," *Proceedings of the Conference on Stormwater Detention Facilities,* American Society of Civil Engineers, New York, 1982.

URBONAS, B. R., AND GLIDDEN, M. W., "Potential Effectiveness of Detention Policies," *Proceedings of the Second Southwest Regional Symposium on Urban Stormwater Management,* Texas A&M University, November, 1984.

U.S. BUREAU OF RECLAMATION, *Hydraulic Design of Stilling Basins and Energy Dissipators,* United States Department of Interior, Bureau of Reclamation, GPO, Washington, D.C., 1964.

U.S. BUREAU OF RECLAMATION, *Design of Small Dams,* United States Department of Interior, Bureau of Reclamation, GPO, Denver, Colo., 1973.

U.S. BUREAU OF RECLAMATION, *Design of Small Canal Structures,* United States Department of Interior, Bureau of Reclamation, GPO, Denver, Colo., 1974.

5

Concrete Basins

5.1 GENERAL

Among the different types of stormwater storage facilities, concrete basins offer greatest flexibility. Due to their structural nature, concrete basins can be configured into almost any geometric shape. Their main advantage is that their sides can be made near-vertical or vertical, which means that right-of-way can be minimized. Their disadvantages include poor aesthetics, high construction cost, and safety. Aesthetics and safety needs can be addressed adequately in a manner similar to what was done in the facility illustrated in Figure 4.2. Often, a concrete basin can be located in out-of-sight locations such as an industrial plant, storage yard, etc. where aesthetic concerns and safety to the general public is mitigated by the limited access to such facilities. As an alternative, concrete storage basins can be turned into underground storage vaults when site conditions so dictate.

Like most other types of storage facilities, concrete basins can be connected both in in-line and off-line to the drainage conveyance system. How they are used depends on the design objectives. Although concrete basins can be open on the top, they are most often used as buried vaults. To date, their greatest use has been in Europe for the control of combined sewer overflows. The remainder of this section describes various flow regulating configurations that may be used in the capture, storage, and treatment of combined waste-water/stormwater flows and, for some of them, separate stormwater runoff systems.

5.2 SYSTEMATIZATION OF STORAGE BASINS

Concrete storage basins can be classified by how they are connected to the storm sewer into two main groups:

- In-line storage;
- Off-line storage.

5.2.1 In-line Storage

An in-line storage basin is connected in series to the sewer. The basin is equipped with an outlet that has less hydraulic capacity than the inlet. Flows pass through the basin undetained until the inflow rate exceeds the outlet's capacity. The excess inflow is then stored within the basin until the basin is full or the inflow rate decreases.

Figure 5.1 shows a schematic representation of an in-line storage facility equipped with a spillway, which operates only when the storage volume of the basin is exceeded. The amount of treatment the inflow receives is determined, among other things, by the holding time of the water in the basin.

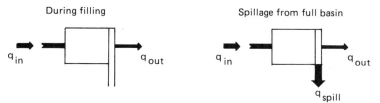

Figure 5.1   In-line storage with a spillway. *(After ATV, 1977.)*

The advantages of in-line storage with a spillway include the following:

- one spillway;
- simple piping arrangement;
- floatables can be skimmed off at spillway;
- storage can be drained by gravity; and
- flexibility in design.

The disadvantages include:

- wide variations in outflow by gravity, and
- difficult to design to be self-cleansing.

Figure 5.2 illustrates an in-line storage where the excess inflow bypasses the basin. When the basin is full, the water in the basin backs up into an upstream splitter and no additional flow can enter the storage basin. In this way, all of the smaller storms and the front end of the larger storms are intercepted. As a result, the smaller and more frequent storm runoff receives treatment. For the larger storms, the more polluted "first flush" will receive treatment, while the remainder of the runoff receives no treatment.

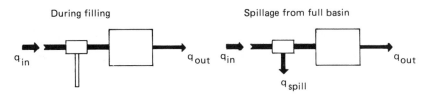

Figure 5.2   In-line storage with an upstream bypass. *(After ATV, 1977.)*

> The advantages of in-line storage with upstream bypass include the following:
> - only one, simple, splitter;
> - simple piping arrangement;
> - emptying of storage by gravity is likely; and
> - greater flexibility in final design.
>
> The disadvantages include:
> - if storage is emptied by gravity, there is a large variation in flow;
> - floatables will enter the storage basin; and
> - difficult to design storage for self-cleansing.

5.2.2 Off-line Storage

An off-line storage basin is connected in parallel to the sewer pipe, whereby the dry weather flow bypasses the storage basin. During a storm, the

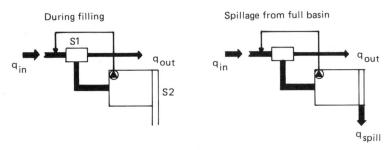

Figure 5.3   Off-line storage with a spillway. *(After ATV, 1977.)*

The advantages of off-line storage with a spillway include the following:
- small head loss in the parallel pipe;
- flow in downstream storm sewer varies less than in a series arrangement;
- floatables can be skimmed off at the spillway; and
- the basin is dry during dry weather periods.

The disadvantages include:
- pipe layout is more complex than in series connection; and
- pumping of the basin is often required to empty.

flow depth in the sewer increases until it overflows a side channel spillway at $S1$ and it spills into the storage basin. Such an arrangement is illustrated in Figure 5.3, which also includes an overflow spillway at $S2$. Flows exceeding the storage basin's capacity spill out of the basin via this spillway.

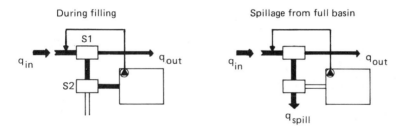

Figure 5.4   Off-line storage basin without a spillway. *(After ATV, 1977.)*

The advantages of off-line storage without a spillway include the following:
- minimum head loss in the bypass sewer;
- less flow variation in the bypass sewer;
- no water in the basin during dry weather flow periods.

The disadvantages include:
- more complex piping arrangement is required;
- will probably require pumping to empty;
- floatables likely to spill out of the system; and
- hydraulics of side channel spillways is complicated.

Figure 5.4 illustrates an off-line storage basin without a spillway. During storm runoff, when the flow exceeds a preselected level, it spills over a side channel spillway $S1$ into the storage basin. The inflow into the storage facility continues until the storage basin is full, or the flow level at $S1$ drops below the side channel spillway's crest. When the basin is full, the water level in the basin is at the same elevation as in the bypass pipe. As a result, the inflow can no longer enter the basin and bypasses it through $S2$, usually receiving no treatment.

5.2.3 In-line and Off-line Combinations

The in-line and the off-line storage arrangements can be combined to work together and function as a single unit. Here, one basin functions as an in-line basin and the other as an off-line basin. By combining these into a composite unit, the advantages of both can be utilized.

Figure 5.5 shows a composite unit connected in series with the sewer system. Under dry weather conditions and during small storms, only the in-line unit is utilized. During larger storms, the in-line basin fills to capacity and the excess flow is diverted through $S1$ into the off-line basin. When both basins are full, any additional runoff then spills through $S2$ out of the second basin.

The idea behind this somewhat complex arrangement is to trap the strongest polluted stormwater in the first basin and to provide some treatment for the remainder of storm runoff. Even in the case where the capacity of the second basin is exceeded, some purification of the stormwater is expected to occur by virtue of it being detained in storage.

An arrangement utilizing a composite unit connected in parallel is illustrated in Figure 5.6. In this arrangement, the dry weather flow bypasses both storage basins. When the flow reaches a preset rate, the excess is diverted into the first basin through a side channel spillway, $S1$. The first basin does not have a spillway, and when it is full, water backs up and is diverted to the second basin through $S2$. When the second basin fills up, the excess flow then spills out of the system through $S3$.

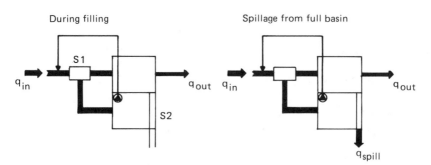

Figure 5.5   Combination storage connected in series.

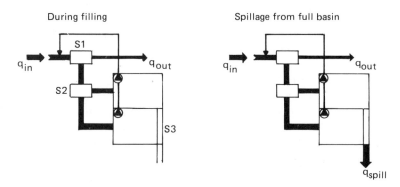

Figure 5.6 Combination storage connected in parallel.

The advantages of the parallel connection are very similar to the ones for the series arrangement, except the water stored in the first basin is assured treatment in a downstream treatment plant. One added benefit is that a more uniform flow rate to the plant is achieved. In areas where the first flush is demonstrated to contain the strongest concentrations of pollutants, the composite arrangement, whether connected in series or parallel, can achieve greater treatment efficiencies than a single storage system. However, this improved treatment is achieved by having a more complex and possibly more expensive installation.

5.2.4 In-line Or Off-line Storage?

The arrangements described in this section are primarily intended for use in a combined sewer system with a wastewater treatment plant located downstream in the system. Also, most of the storage arrangements described so far can also be used for flow equalization. As such, they can reduce flow peaks to match the conveyance or treatment capacity of the downstream system.

Some treatment of separate stormwater runoff can be achieved, however, without additional downstream treatment facilities if an in-line system is employed. When storage is the only source of treatment, the operation and maintenance activities will increase, consistent with the treatment levels.

In a combined sewer system, whether an in-line or an off-line facility should be used will depend on whether the system will experience a strong first flush. In combined sewers and urban areas with flat terrain, the accumulation of pollutants in storm sewers may be the prime reason for the observed first flush. Munz (1977) observed that the first flush is strongest if the runoff flow time is less than 10 minutes. Also, according to Munz (1977), the most pronounced flushing effect in a storm sewer occurs when pipes have dry weather flow velocities between 1.5 and 2.5 feet per second. When the velocities are higher, very little accumulation of pollutants occurs in the storm sewer. With lower velocities, the flushing process appears to be inefficient, and flushing is not limited to the beginning of the runoff.

Also, more significantly polluted first flush is generally associated with smaller watersheds, where the mixing of flows from large areas is not present.

First flush in separate storm sewers is not always observable. As more and more data, such as that collected during the Nationwide Urban Runoff Program (EPA, 1983) become available, it seems that a strongly polluted first flush may be found in some urban centers and not in others.

Beside trapping the first flush of storm runoff, the selection of the storage system needs to consider the capacity of the downstream conveyance system, the presence and capacity of a treatment plant, and the size and nature of the watershed downstream of the storage. For instance, if the watershed between the storage facility and the treatment plant contributes much storm-water, the pollutants are best trapped and held in a storage facility which is connected in parallel. If this is not done, the effects of detention will be lost through dilution and possible spilling into the downstream system. Table 5.1 compares the conditions for the selection of a storage system arrangement for the purpose of trapping the first flush.

5.3 TECHNICAL CONFIGURATION

The physical configuration of a concrete storage basin is largely determined by site conditions. For instance, the vertical fall in the sewer system will determine if the basins will be drained by gravity or if pumping will have to be used.

TABLE 5.1 Conditions for Trapping of the First Flush

Flow Time in the Watershed	
<10 minutes	good
10–20 minutes	average
>20 minutes	poor

Dry Weather Flow Velocity in Sewer	
<0.5 meters per second	average
0.5–0.8 meters per second	good
>0.8 meters per second	average

Watershed Condition Downstream of Storage	
No addition of runoff downstream	good
Addition of stormwater downstream:	
Parallel connection	average
Series connection	poor

After Munz, 1977.

Also, site geometry will dictate how the installation is configured in the horizontal plan. This can be especially critical in dense urban areas where right-of-way is very limited. Fortunately, concrete storage tanks can be easily configured into any desired horizontal plan.

Concrete storage basins can be built with an open or a closed top; however, the closed basin is used almost exclusively in densely urbanized areas. A closed basin has advantages for safety and odor control, especially in combined sewer systems. Only at the treatment plant, or in isolated and fully fenced areas, is an open basin a viable option in a combined sewer system.

5.3.1 Vertical Considerations

Refer to Figure 5.7 and examine some of the more important aspects of a storage basin's vertical configuration.

According to Koral and Saatci (1976), there are certain height restrictions for a cost-effective installation of concrete detention vaults. For example, the elevation difference between the outlet and the maximum storage level should be 7.5 to 11.5 feet, where the lesser depth is intended for smaller storage basins. If the suggested depths are not achieved, the basins require excessive horizontal area, are more expensive to build, have less efficient flow conditions, have limited accessibility for cleaning, etc. Figure 5.8 shows the optimum vertical dimensions that were recommended by Koral and Saatci (1976).

5.3.2 Horizontal Plan

Obviously, the area that the storage facility will occupy will depend, among other things, on the height limitations and right-of-way availability. A rectangular shape offers certain cost and maintenance advantages. Also, experience suggests that flat-bottomed basins should have their width equal to one-half to two-thirds of the length of the basin. For basins with sloping parallel grooves in the bottom, the length is determined by the distance over

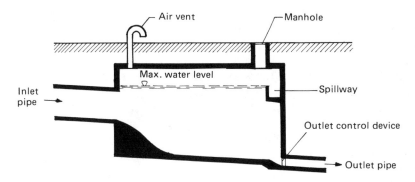

Figure 5.7   Vertical configuration of a concrete basin. *(After ATV, 1977.)*

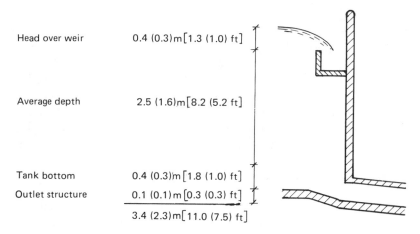

Head over weir	0.4 (0.3)m[1.3 (1.0) ft]
Average depth	2.5 (1.6)m[8.2 (5.2 ft]
Tank bottom	0.4 (0.3)m[1.8 (1.0) ft]
Outlet structure	0.1 (0.1)m[0.3 (0.3) ft]
	3.4 (2.3)m[11.0 (7.5) ft]

Figure 5.8 Minimum vertical dimensions for concrete detention tanks. *(After Koral and Saatci, 1976.)*

which the grooves can be effectively flushed clean. This is further discussed in section 5.3.5.

Site availability, however, will sometimes dictate a variation from a standard rectangular plan. It may be necessary to design irregularly shaped basins, especially in densely populated areas. In such cases, you can expect construction and basin cleaning costs to be higher.

Round and octagonal basins can be more expensive to build; however, they offer self-cleansing possibilities often lacking in other configurations. As illustrated in Figure 5.9, by arranging the inflow tangentially and the outlet at the center, the tank can be made virtually self-cleansing.

5.3.3 Inlet Pipe

The manner by which wastewater is fed into a storage basin depends on whether the basin is in series or in parallel with the sewer. When in series, all of the stormwater enters the storage basin. As a result, no separate inlet structures are needed. To gain additional storage capacity, it is advantageous to let the water rise in the basin until it backs up into the inlet. If this is the intended design, however, be sure that the water does not back up into nearby basements. When there is a risk of water backing up into basements, investigate the possibility of collecting the wastewater from affected basements in a separate sanitary sewer that bypasses the storage basin.

When the basin is filled by two or more sewers coming from different directions, combine the sewers into a single inlet pipe. If, however, multiple inlets are necessary, then pay particular attention to the flow patterns in the basin to insure that no unusual "dead zones" or adverse flow conditions are created.

When the storage basin is connected in parallel, the inflow takes place through a separate inlet structure. This structure consists of a side channel overflow or spillway designed to function only after a predetermined flow in the sewer is exceeded. Figure 5.10 shows one schematic example of a side channel overflow. For a more detailed description of the analysis and design of side channel spill structures, refer to advanced open channel design textbooks or special literature.

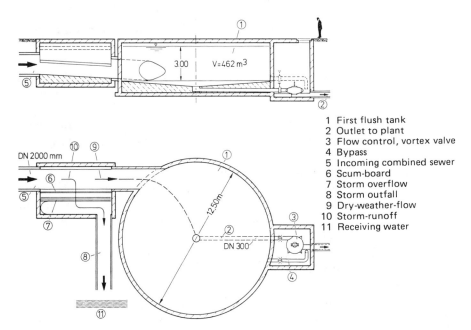

1 First flush tank
2 Outlet to plant
3 Flow control, vortex valve
4 Bypass
5 Incoming combined sewer
6 Scum-board
7 Storm overflow
8 Storm outfall
9 Dry-weather-flow
10 Storm-runoff
11 Receiving water

Figure 5.9   Round concrete basin. *(By Umwelt and Fluidtechnik GmbH, West Germany.)*

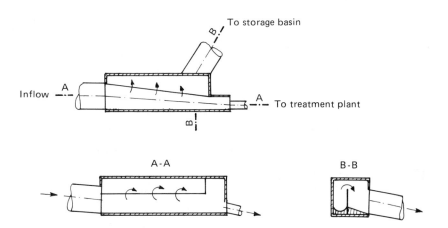

Figure 5.10   An example of a side channel overflow.

5.3.4 Inflow Arrangements

Before designing an inflow structure for a storage basin, determine if

1. the basin is to provide removal of solids before the excess flow is permitted to spill to the receiving waters, or
2. the basin is going to be without a spillway.

In the first case, the inflow must be designed to enter the basin without resuspending the settled solids. This will require energy dissipation and rapid diffusion of the inflow plume. Figure 5.11 contains an example of how the inflow plume can be diffused quickly through the use of baffles. When designing such baffles, be careful not to create a rotating flow motion within the basin itself.

For basins without spillways, settling of solids on the basin floor should be avoided. In fact, the resuspension of solids into the water column facilitates the cleaning of the storage basin. To achieve this, the inlet is designed to make the installation as self-cleansing as possible.

To keep the sediments suspended, the energy of the inflow water can be used to clean the basin's floor. This can be done by arranging the inlet(s) to create as much circular flow and turbulence in the basin as possible. First, try to maintain the inlet pipe elevation as high as possible so the water can accelerate into the basin. Next, locate the inlet(s) and the outlet tangentially to maintain rotational flow in the basin, as in Figure 5.12.

To preserve kinetic energy of the inflow so that it can be transferred to rotational flow, the inlet can be equipped with a parabolic profile. The water is fed into the basin by a groove in the upstream end of the basin (see Figure 5.13) or by a sloping inlet pipe, as in Figure 5.14.

5.3.5 Basin Bottom Configuration

One has to expect that some sand, silts, and sludge will settle out in the basin. To expedite cleaning, the bottom must be carefully designed and constructed. Typically, we find three basic configurations that help to achieve this:

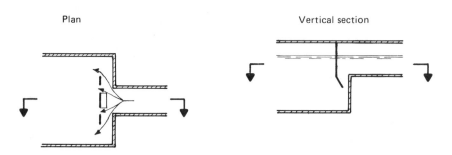

Figure 5.11   Baffle arrangement at the inlet. *(After Koral and Saatci, 1976.)*

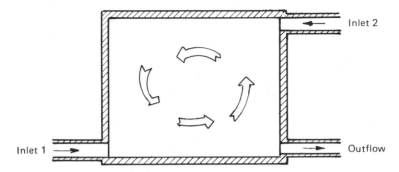

Figure 5.12   Placement of inlets and outlet to promote rotational flow in basin. *(After Malpricht, 1973.)*

Figure 5.13   Example of inlet groove in a storage basin.

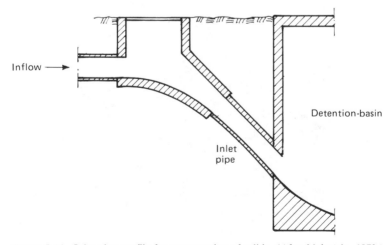

Figure 5.14   Inlet pipe profile for resuspension of solids. *(After Malpricht, 1973.)*

- flat bottoms,
- bottoms with multiple parallel grooves, and
- bottoms with a single continuous groove.

Flat Bottoms.   Of all the basin bottom configurations, the flat bottom is, by far, the simplest and least costly to build. Flat bottoms also are easiest to clean and best for access to all parts of the basin. On the other hand, flat bottoms do not lend themselves as readily as the other two types to automated cleaning. Figure 5.15 shows a concrete storage basin with a flat bottom.

To permit easy access to all parts of the basin for maintenance, the bottom should be sloped no more than 10%. The lower limit for this slope is 3% which is needed for good drainage of the basin floor.

Basins connected in series with the sewer are usually equipped with a bottom groove (i.e., trickle flow channel) to convey the dry weather flows. The hydraulic capacity of this groove has to be in balance with the outlet capacity. Only when the outlet capacity is exceeded, the water should spill out of the groove onto the basin floor. A low flow groove can also be used in basins connected in parallel to the storm sewer. Even in a parallel arrangement, it helps to guide the low flow and the sediments it carries directly to the outlet.

A simple way to build a low flow groove is to use sanitary sewer pipe halves, as shown in Figures 5.16 and 5.17. These pipe halves are grouted into a precast slot on the basin floor, taking care not to create sharp edges or corners. The groove itself may be located in the center or at one of the edges of the basin, as shown in Figure 5.18.

Round and octagonal basins are, almost without exception, constructed with a flat, cone-shaped bottom. Here, the bottom slopes toward the center, where an outlet is located. The dry weather flow follows a parabolic path, as illustrated in Figure 5.9 to the central outlet. This arrangement provides little interference with the circular flow within the basin.

Figure 5.15 Concrete basin with a flat bottom.

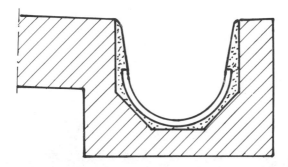

Figure 5.16   Low flow groove configuration. *(After Malpricht, 1973.)*

Figure 5.17   Bottom groove constructed of pipe halves.

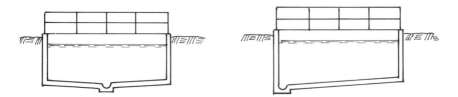

Figure 5.18   Possible low flow groove locations.

Bottoms With Parallel Grooves.   To facilitate automatic cleaning, the basin's bottom is sometimes equipped with a number of parallel longitudinal grooves. Each of the grooves is cleaned separately by the use of flushing water. Clearly, this bottom design is more expensive to build than a flat bottom. In addition, it is more difficult for maintenance personnel to move around a grooved bottom. An example of a basin with parallel longitudinal grooves is shown in Figure 5.19. Three different configurations of groove design are illustrated in Figure 5.20.

Figure 5.19 Bottom with parallel longitudinal grooves.

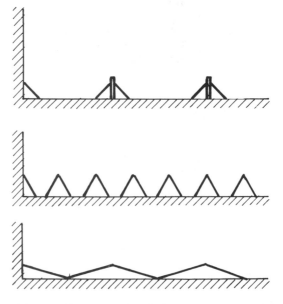

Figure 5.20 Three configurations of basin bottom grooves. *(After Koral and Saatci, 1976.)*

Originally, the purpose for the installation of parallel grooves in a storage basin floor was to make the bottom self-cleansing as it empties. Experience has shown, however, that the stored water does not accelerate sufficiently to achieve adequate velocities to flush away the deposits. As a result, the bottom has to be cleaned occasionally during the dry weather periods. Cleaning is accomplished by flushing through the outlet using sediment-free water, usually one groove at a time. For this reason, the grooves need to slope at least 3% to 5% toward the outlet; the steeper the slope, the more effective the flushing.

Bottoms With a Single Continuous Groove.   If properly designed, a basin with a single continuous groove can be totally self-flushing during the dry weather flow periods. In this design, the bottom has one continuously sloping groove between the inlet and the outlet. Figure 5.21 illustrates this concept, while Figure 5.22 is a photograph of an actual installation. This concept was introduced in Switzerland by Koral and Saatci (1976).

To achieve reasonable flushing velocities, the longitudinal slope has to be no less than 2%. The slope has to be increased in each U-turn to compensate for the energy loss at each bend. This means that the total vertical fall between the inlet and the outlet can become quite large. As a result, this type of a bottom is practical only for relatively small basins with volumes not exceeding 20,000 cubic feet.

Figure 5.23 illustrates examples of some common groove configurations. Basins with this type of bottom are more costly to build than basins with flat bottoms. To reduce costs, prefabricated groove elements have been introduced in Switzerland. These elements can be incorporated into designs as an alternate, and the bidding process can demonstrate which construction technique is most economical at any given site.

Based on the experience gained in Switzerland, Koral and Saatci (1976) compiled the following set of practical recommendations for the design of basins equipped with a continuous low flow groove:

- Dry weather flow velocity in the groove should be no less than 0.7 meters per second (2.3 feet per second).
- The depth of dry weather flow should be no less than 3 centimeters (1.2 inches).
- Maintain side slopes of the groove between 1:1 and 1.4:1.
- The head loss at each 180 degree turn is estimated at 1 to 2 centimeters (0.4 to 0.8 inches).

5.3.6 Outlet Structure

The outlet structure, in combination with the storage capacity, affects the size and nature of downstream facilities. Namely, the size of sewers, pumps, treatment tanks, etc. depend, to a large degree, on how the outlet and storage are configured.

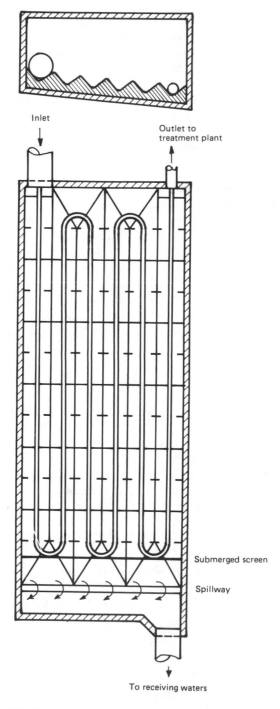

Figure 5.21 Bottom design with a single continuous groove. *(After Koral and Saatci, 1976.)*

Figure 5.22   Storage basin with a single continuous groove.

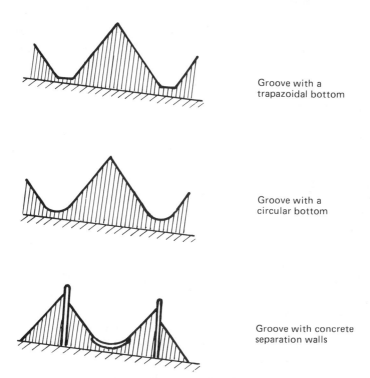

Groove with a
trapazoidal bottom

Groove with a
circular bottom

Groove with concrete
separation walls

Figure 5.23   Examples of basin bottom groove. *(After Koral and Saacti, 1976.)*

Often, it is desirable to maintain the outflow constant. This is not easy to accomplish, since most outlet controls exhibit a linear logarithmic relationship between the headwater depth and the discharge. Even when the basin is emptied by a pump, the pumping efficiency can vary with the submergence of the pump.

To reduce sediment buildup on the basin floor, it is advisable to reduce the period of time the water is out of the bottom groove and on the basin floor. Unnecessary shallow ponding behind the outlet also increases sediment deposition and the maintenance frequency of the facility. Depressing the outlet into the basin floor approximately 4 to 8 inches (see Figure 5.8) can significantly reduce these types of problems.

The outflow from a storage basin can be controlled in several ways, including the use of:

- a fixed outlet orifice or nozzle,
- a choked outlet pipe,
- adjustable gates,
- pumps, and
- special flow regulators.

Because the outlet plays such a significant role in the detention process, it deserves special attention. As a result, Part 2 of this book is dedicated to this topic.

5.3.7 Spillways and Emergency Outlets

In addition to the operational outlet, concrete storage basins are typically equipped with an emergency outlet, which begins to operate when the basin is full. An emergency outlet can be a large pipe or a spillway. If the downstream sewer has excess capacity, the spilled water can be routed to it; otherwise the spilled water is conveyed, untreated, to the receiving waters.

To prevent floatables from discharging over the spillway, the spillway can be equipped with a skimmer. The skimmer is a separate structure located just upstream of the spillway, as shown in Figure 5.24.

The spillway in the storage basin should be located at the side opposite from the outlet, because the floatables often collect near the outlet. By keeping the spillway away from the outlet, the chances for floatables rising between the skimmer and the weir as the basin fills are reduced.

The emergency outlet has to be able to handle all of the incoming flow. Thus, the spillway is sized to pass the maximum inflow, without taking into account the capacity of the main outlet. The spillway's design capacity has to anticipate the possibility that the main outlet may become clogged or may have to be taken out of operation for system maintenance.

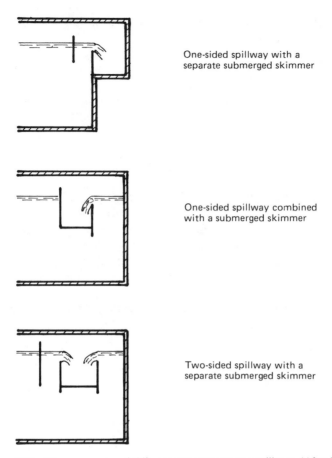

One-sided spillway with a
separate submerged skimmer

One-sided spillway combined
with a submerged skimmer

Two-sided spillway with a
separate submerged skimmer

Figure 5.24   Three examples of skimmer arrangement at a spillway. *(After Koral and Saacti, 1976.)*

5.4 SUPPLEMENTAL PROVISIONS

When installing underground detention storage, provisions need to be made for the installation of electricity, ventilation, and cleaning water. These are needed for the operation, cleaning, and maintenance of the storage basin and its appurtenances.

5.4.1 Electrical Equipment

Electric power is needed for lighting and for the operation of pumps, gates, and other mechanical equipment. All electrical equipment has to be corrosion resistant, flood-proof, and explosion-proof. The latter is because methane gases have been shown to accumulate inside an underground storage basin in combined storm sewer systems.

Light fixtures inside the basin should be located to permit easy access for replacement. Light fixtures on walls are easier to access than on the ceiling. The circuit breakers and switches should, however, be located outside of the detention basin vault. When possible, all electrical equipment, except the lights, should be located in a special ventilated and heated room. In small installations, this can be a freestanding switch box as shown in Figure 5.25.

5.4.2 Ventilation

It is very important to provide effective ventilation for underground detention. Experience suggests that good ventilation is accomplished when four- to six-fold exchanges of air per hour are achieved. Although the inflow and outflow pipes can provide some ventilation of the basin, their contribution is unreliable and should not be considered in the design. Also, the ventilation openings should be designed to prevent air from being trapped between the basin ceiling and the water surface.

The size of the vent openings will depend on the basin location, wind conditions, and ventilating arrangement. To achieve the desired air exchange rate, the openings should be sized for the following maximum air flow velocities:

0.25 m/sec in wind-sheltered locations,

0.50 m/sec in reasonably windy locations, and

0.75 m/sec when a chimney effect is provided.

Figure 5.25   Freestanding electrical switch box.

5.4.3 Cleaning Water

It is advisable to make provisions for fresh water to wash down the storage basin walls, skimmers, weirs, etc. and to flush out the deposits that accumulate on the bottom. The most common sources for this water are

- municipal water systems,
- temporary connection to fire hydrants,
- pumping from a nearby stream, pond, or lake,
- water wells, and
- water trucks.

The selection of the water supply will depend on what is most readily available or practical. For large installations, permanent water supply installations are preferable. An internal water distribution system is handy and provides easy access to all parts of the basin, as seen in Figure 5.26. When connecting to a water system, be sure to install backflow prevention valves with antisiphon devices. Backflow into a public or domestic water supply system is not acceptable and should never be permitted to occur. Be sure to also provide antisiphon devices when using nearby fire hydrants, water wells, streams, lakes, etc. as a source of cleaning water.

5.5 OPERATION AND MAINTENANCE

The concrete storage basin, especially if it is underground, will be subjected to extremely harsh operating conditions. It will be subjected to

Figure 5.26   Cleaning water system in a basin.

- high humidity,
- organic sludge deposits,
- corrosive gases,
- intermittent operation, and
- microbe and fungal attack.

Some of the operational problems can be mitigated by appropriate design. However, regular inspection and maintenance have to be provided if satisfactory function is to be maintained. The designer can facilitate maintenance, however, by incorporating labor-saving devices into the installation.

5.5.1 Manholes And Access Openings

All covered storage basins need access openings for maintenance personnel and equipment. For larger basins, a permanent stairway, as seen in Figure 5.27, can be provided. For smaller installations, a permanently installed ladder can be used, but that option is not as convenient, or as safe, as a stairway. If possible, personnel access should be from an above-ground building, as in Figure 5.28, which can also house all of the electrical controls and the valves for the cleaning water system.

Access openings should also be provided for the purpose of moving cleaning and maintenance equipment and materials in and out of the basin. One of the openings should be located directly above the outlet for cleaning when the basin is full and the outlet is clogged.

In addition to providing maintenance access, the same openings can also

Figure 5.27   Access stairway for maintenance personnel.

Figure 5.28   Access building containing electric controls.

be used for ventilation and to admit daylight into the basin. When the openings permit sufficient daylight into the basin, electric lighting may not be necessary.

5.5.2 Inspection Access

It is advisable to provide separate inspection walkways inside large storage basins. These make the inspection of a basin more comfortable and facilitate more frequent visits. Two examples of how such walkways can be arranged inside of a basin are shown in Figure 5.29. To reduce costs, the use of the emergency spillway for an inspection walkway is possible. However, access will not be feasible when the spillway operates. An example of this type of an arrangement is shown in Figure 5.30.

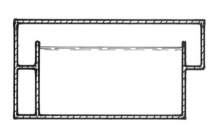 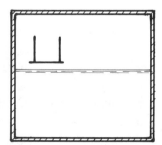

Figure 5.29   Two examples of inspection walkways. *(After Koral and Saatci, 1976.)*

Figure 5.30 Emergency spillway as an inspection walkway.

5.5.3 Regular Inspection

A routine inspection schedule is a precondition for satisfactory operation of the storage basin. The frequency of inspection will vary with the complexity of the installation and whether or not it contains mechanical equipment. In addition, inspection of the basin after each storm will reveal if the cleaning of clogged outlets or the removal of unusual accumulations of deposits is needed. Keeping a log of all inspections including dates, times, names of inspectors, and findings can help identify long-term trends and prevent major problems.

The inspector should attend to the following during each visit:

- Check electrical parts.
- Test operation of pumps.
- Check outlets for clogging.
- Check for sludge deposits.
- Inspect the water distribution system.
- Examine measuring devices.
- Look for excessive condensation.
- Look for corrosion damage.
- Look for signs of early damage or deterioration.

5.5.4 Cleaning of Basins

Because there will be sedimentation inside a concrete storage basin, it will need occasional cleaning. Proper design can facilitate this task and, in some cases, automate it. Nevertheless, supplemental manual cleaning will always be needed regardless of the design used.

Flushing of Flat Bottoms.   Whenever the storage basins with flat bottoms have to be flushed with water, proper design of the installation will significantly ease this task. Provide a water supply with sufficient pressure to do the job. Also, the bottom has to be properly sloped, as described earlier.

Flushing efficiency is directly related to the energy of the available water. The pressure in a municipal water system can reach 80 pounds per square inch. However, it often is significantly lower, and in those cases it will be necessary to install a booster pump. The pump is usually connected to a flushing water tank, as illustrated in Figure 5.31. Such a tank insures that the municipal water system will not be overloaded when the pump operates. Also, the air gap between the water inlet and the stored water insures that the stored water will not siphon back into the municipal system.

Flushing Bottoms with Parallel Grooves.   Flushing of grooves is achieved by running a large quantity of water in a short period of time. This rapid application of water carries away the accumulated deposits. Obviously, the longer the groove, the more water will be needed, so be sure to account for this during design. Both clean water or stormwater can be used for flushing of the grooves.

To provide this shock load of flushing water, the storage basin needs to be equipped with special flushing tanks, which are designed to feed the water

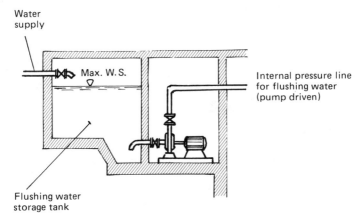

Figure 5.31   Flushing water storage tank. *(After Malpricht, 1973.)*

to each groove separately. The outflow rate is regulated with gates or by the tilting of the tank. These storage tanks can be either fixed or movable; the latter is illustrated in Figure 5.32.

It is also possible in some cases to use the inlet pipe to store flushing water. This is done by temporarily damming up the pipe and then rapidly releasing the water. However, when using this technique, be sure that the backed-up water does not enter upstream basements.

Cleaning with Scrapers.   When flushing with water will not remove the deposits, the use of mechanical scrapers can get the job done. This solution is expensive and is only viable when the bottom configuration can accept such equipment. The use of bottom scrapers is most common at pumping stations and at sewage treatment plants. (See Figure 5.33.)

Cleaning with Mobile Cleaning Equipment.   In certain cases, attempts have been made to clean large storage basins having flat bottoms with small

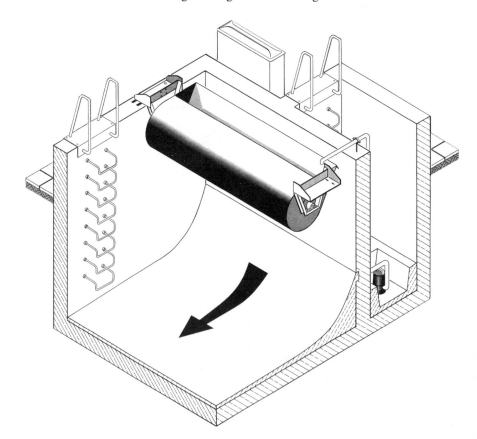

Figure 5.32   A movable flushing water container. *(By Umvelt und Fluidtechnik GmbH, West Germany.)*

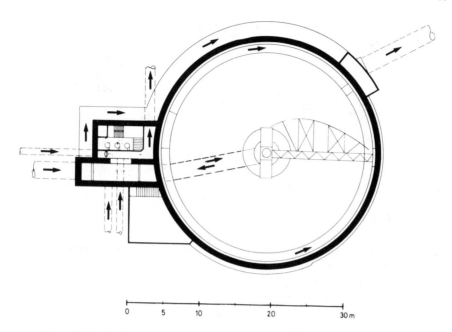

Figure 5.33   Storage basin with a mechanical bottom scraper. *(After Larsen, 1978.)*

mobile cleaning units. However, this technique has not yet gained wide acceptance for the cleaning of underground basins. It is, however, in wide use for the cleaning of large, above-ground installations.

5.5.5 Water Level Record

It is essential to record water levels in the basin to provide an operational record and future design and maintenance information. This can be done by installing a water level recorder. To calibrate and to verify the operation of the recording devices, a graduated measuring rod should be installed in a location that is clearly visible and accessible when the basin is full.

REFERENCES

ATV, *Guidelines For Dimensions, Configuration and Operation of Rain Detention Basins,* Arbeitsblatt A 117, 1977. (In German)

ATV, *Guidelines For Dimensions and Configuration of Rain Relief Installations in Mixed Water Canals,* Arbeitsblatt A 128, Second Edition, 1978. (In German)

BAYERISCHES LANDESAMT FUR WASSERWIRTSCHAFT, *Planning Aids For the Configuration of Rain Overflow Basins,* Munich, 1978. (In German)

BROMBACH, H., "Model Studies of the Self-Cleaning Behavior of Rain Overflow Basins," Wasser und Boden No. 2, 1979. (In German)

DODSON, K., AND LINDBLOM, *Evaluation of Storm Standby Tanks*, Columbus, Ohio, EPA 11020 FAL 03/71, 1971.

EIDGENOSSISCHES AMT FUR UMWELTSCHUTZ, *High Water Relief and Rain Overflow Basins*, 1977. (In German)

EPA, *Results of the Nationwide Urban Runoff Program*, Final Report, U.S. Environmental Protection Agency, 1983.

HEDLEY, D., AND LOCKLEY, J. C., "Use of Retention Tanks on Sewerage Systems: A Five Year Assessment," Water Pollution Control, 1978.

INTERNATIONALE GEWASSERSCHUTZKOMMISION FUR DEN BODENSEE, *Rain Relief Installations, Dimensions and Configuration*, Report No. 14, 1973. (In German)

KALINKA, G., "Use and Experience With Cleaning Equipment for Rain Basins," Wasser und Boden No. 11, 1979. (In German)

KORAL, J., AND SAATCI, C., "Self-Cleaning Rain Overflow Basins With Snake Grooves," Wasserwirthschaft No. 10, 1974. (In German)

KORAL, J., AND SAATCI, C., *Rain Overflow and Rain Detention Basins*, Second Edition, Zurich, 1976. (In German)

KRAUTH, K., "Recipient Relief by Treating Rain Water in Rain Water Overflow Basins," Wasserwirtschaft No. 2, 1973. (In German)

KROPF, A. "Detention Basins And Water Clarification Plants," Scheizerische Bauzeitung No. 18. 1957. (In German)

LARSEN, E., "Open and Closed Basins in Combined Systems," Seminar om Utjevningsbassenger, Marsta, 1978. (In Danish)

MALPRICHT, E. Construction and Operation of Dentention Basins," Korrespondenz Abwasser No. 5, 1973. (In German)

MASSACHUSETTS, COMMONWEALTH OF, METROPOLITAN DISTRICT COMMISSION, *Cottage Farm Combined Sewer Detention and Chlorination Station*, Cambridge, Mass., EPA-600/2-77-46, 1976.

MILWAUKEE, CITY OF, AND CONSOER TOWNSEND AND ASSOCIATES, *Detention Tank For Combined Sewer Overflow, Demonstration Project*, Milwaukee, Wis., EPA-600/2-75-071, 1975.

MUNZ. W., "Dimensions of Storm Water Basins," Gas-Wasser-Abwasser No. 3, 1975. (In German)

MUNZ, W., "Storm Basins and Stormwater Detention," Wasser/Abwasser No. 1, 1974. (In German)

MUNZ, W., "Storm Basins and Storm Water Detention," Wasser/Abwasser No. 9 and 11, 1973. (In German)

MUNZ, W., "Storm Water Overflows With and Without Detention Basins," Eidg. Anstalt fur Wasserversorgung, Abwasserreinigung & Gewasserschutz, Publ. No. 645, Zurich 1977. (In German)

NASSAU, K., "Stormwater Overflow Basins; Construction, Operation and Cost Aspects," Gwf-Wasser/Abwasser No. 2, 1978. (In German)

UMVELT UND FLUIDTECHNIK GMBH, Bad Mergenlheim, West Germany, Product Information.

6

Storage in Sewer Networks

6.1 GENERAL

Some of the existing combined wastewater-stormwater sewers are large enough to handle runoff from relatively large intensity storms without a surcharge in the system. As an example, a sewer designed for a five-year storm is expected to reach, or exceed, its capacity on the average once every five years. For storms that produce less runoff than the design storm and may account for more than 95% of all of the rainstorms, the combined sewer has excess capacity. This excess can be used to temporarily detain storm runoff.

To utilize this excess capacity for detention, real time operation of the system is required. This is done through the use of flow regulators, rainfall measurement, and flow sensors in combination with the predictions of runoff as the storm is occurring. Torno et. al. (1985) compiled several papers that discuss the state-of-the-art technology of real time control in combined storm sewers. This is an excellent reference that should be studied by anyone interested in this topic.

6.2 VARIOUS FLOW AND STORAGE REGULATORS

Utilization of available storage volume in a sewer requires flow regulators inside a storm sewer. In-line regulators permit controlled impoundment of stormwater in the sewer system and may be

- fixed type,
- movable type, or
- patented, special-purpose type.

The *fixed type* regulators include installations that result in a permanent increase in depth or pressure level in the pipe. Examples of this would be a raised spillway crest or a constricted section of the sewer. This is feasible only where storm sewer surcharging is acceptable and will not cause the storm flows to surface above ground or back up into basements.

Movable type of regulators encompass a wide variety of devices that can vary the release rate and the actual volume being stored on an as-needed basis. The regulator is often a remote controlled valve, gate, inflatable weir, etc. that controls the water level or pressure in the sewer. When combined with rainfall sensing and runoff forecasting, movable type regulators can optimize the storage in the sewer network, and insure adequate capacity during larger storms.

The greatest advances that are expected in combined stormwater sewer management technology are predicted by Schilling (1985) to occur in the area of real time control. The advent of inexpensive computerized controls, linked to a network with a central control station in combination with weather forecasting and radar, will offer the potential for flow optimization in the entire network. This is not an easy task, since storm rainfall distribution over an area and the movement of each storm is not always possible to predict. Also, the equipment that is required is complex and, as a result, subject to occasional failures, even under the best maintenance environments.

In recent years, we have seen the emergence of special control devices. Some of these are patented, while others were developed by public agencies. These include devices such as the Steinscrew and the Hydrobrake system. These two are completely self-regulating and can be used in separate storm sewers and combined storm sewers. The technical configuration of the aforementioned various flow regulating devices is discussed in Chapter 11.

6.3 CHANGING THE FLOW ROUTE IN SEWERS

Most sewer systems in existence were designed to carry away stormwater as quickly as possible. This approach can occasionally lead to a "collision" of peak flows at the confluences of various subbasins. If the original design did not anticipate this, then some of the downstream portions of the system may be overloaded.

In some cases, it is possible to balance out the flow peaks by changing the flow routes. This is especially the case if real time control to divert the flows is used. As an example, one can transfer water from one subbasin to another in a fashion that puts the hydrograph peaks out of phase. To accomplish this, it is necessary to install cross-connections between the subbasins, which may or may not be cost-effective.

When planning a new storm sewer system, the travel time of each branch can be optimized to avoid collision of peaks. This is possible through varying the layout and flow lengths of the pipe system. However, detailed storm runoff routing modeling is needed to provide the details of flow in various parts of the system for such a design.

6.4 SOME PRACTICAL OBSERVATIONS

Combined sewer networks in densely urbanized areas generally are the result of many years of growth and expansion. The excess volume that may have been available when the system was new may no longer be available. As the urbanized area expands, the sewer system is modified in response to flow increases and public input (see Figure 6.1).

Because combined sewer systems are sensitive to changes in runoff conditions, especially to the density of urbanization, the use of pipes for detention storage is often not possible. Also, the use of choked sections that back up stormwater in a sewer can cause local flooding. Therefore, when considering the use of storm sewers for detention, detailed and thorough analysis of the system capacity must be performed. Every feasible scenario of the temporal and spatial distribution of rainfall needs to be examined using distributed routing models before deciding if, and how, pipe storage will be utilized.

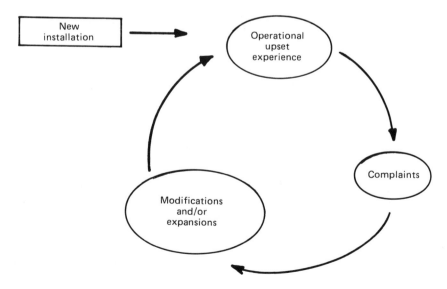

Figure 6.1   Operational experience influence on system modification.

REFERENCES

SCHILLING, W., "Urban Runoff Quality Management by Real-Time Control," *Urban Runoff Pollution*, Springer-Verlag, Berlin, 1986.

TORNO, H., MARSALEK, J., AND DESBORDES, M., *Urban Runoff Pollution*, Springer-Verlag, Berlin, 1986.

7

Pipe Packages

7.1 BASIC CONFIGURATION

Pipe packages is a term describing a detention storage facility consisting of one or more buried large-diameter pipes. These pipes are placed in parallel rows, and each pipe is connected to a common inlet chamber and an outlet chamber. Usually, but not always, pipe packages are connected in series to the sewer pipe. As illustrated in Figure 7.1, the connection of the package to the sewer takes place at the two end points, namely, inlet and outlet chambers.

The size of a pipe package is determined by the storage volume requirements and by the physical space availability at the installation site. The package does not need to be installed in a straight line along its entire length. It can change direction anywhere along its length to fit the site limitations. A typical pipe package is equipped with a flow regulator that is installed in the outlet chamber and an overflow spillway located at either the inlet or the outlet chamber.

This chapter describes the following aspects of a pipe package installation:

- pipes forming the package,
- inlet chamber,

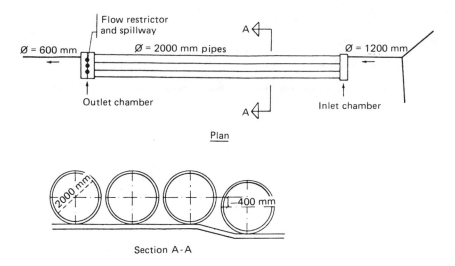

Figure 7.1   Basic layout of a pipe package storage.

- outlet chamber,
- spillway,
- access openings, and
- operation and maintenance.

7.2 PIPES FORMING THE PIPE PACKAGE

The size of the pipes used in a package can vary considerably. However, to facilitate inspection and cleaning, it is recommended that, as a minimum, 54-inch diameter pipes be used. For similar reasons, the pipes should be laid at a minimum slope of 2% to avoid standing pockets of water, which can occur due to lack of precision during construction.

Although sediments will settle out inside pipe packages, this same deposition can be reduced by installing one of the pipes somewhat lower than the others; see Figure 7.1. Confining the low flows to one pipe will help the system to become self-cleansing. To keep the other pipes from filling during low flows, the elevation difference between the low flow pipe and other pipes needs to be set to keep the low flows confined only to the low flow pipe.

Despite the fact that one pipe is set lower to carry the low flows, deposition of sediments and sludge will still occur. These deposits can be cleansed by maintenance personnel by diverting the dry weather flows manually to one pipe at a time. This is done by inserting gates in front of other pipes into frames that were installed for this purpose during construction. Although this system may be somewhat complex, it can facilitate the cleaning of pipe packages in combined sewer systems. This may not be possible in separate storm sewers unless there is a significant dry weather flow in the sewer.

7.3 INLET CHAMBER

At the upstream end, the pipe package is connected to the network through the inlet chamber (see Figure 7.2). This chamber can be constructed of reinforced concrete, or it can be assembled using large pipe fittings. Whichever way it is built, it needs to be large enough to permit comfortable access to all of the pipes by the maintenance personnel and their equipment.

7.4 OUTLET CHAMBER

At the downstream end the pipe package is connected to the storm sewer system through an outlet chamber. This chamber can also be built using reinforced concrete or assembled using large pipe fittings. The flow leaves the outlet chamber through one of the following types of flow regulators:

- a fixed orifice,
- a fixed outlet nozzle,
- a small outlet pipe,
- adjustable gates, or
- special (i.e., patented) flow regulators.

Figure 7.2   Three 63-inch pipes at inlet chamber.

7.5 SPILLWAYS

To prevent water from surfacing at manholes or in upstream basements, emergency overflow spillways need to be installed either at the inlet or the outlet chamber. When runoff from larger storms exceeds the storage capacity of the pipe package, the excess can then be safely spilled, untreated, to the receiving waters.

If possible, locate the spillway at the inlet chamber. This type of an arrangement is illustrated in Figure 7.3, where the water has to back up into the inlet chamber before the spillway operates. As a result, the storage capacity in the pipe package is utilized fully. When locating the spillway inside the inlet chamber, be sure that water cannot back up into the upstream system and cause damages or other problems.

Figure 7.4 illustrates a spillway located in the outlet chamber. There is less risk of causing upstream flooding; however, as illustrated in the figure, it is possible that the storage capacity may not be fully utilized. Also, there is a greater risk in this configuration that the bottom deposits in the pipes will be resuspended and spilled, untreated, into the receiving waters.

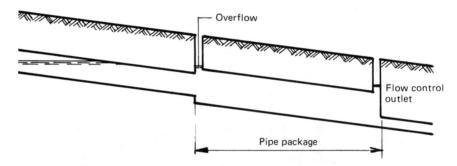

Figure 7.3   Pipe package with spillway at inlet chamber. (*After ATV, 1978.*)

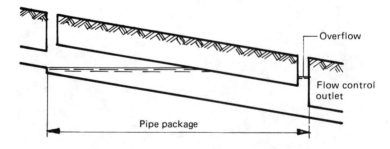

Figure 7.4   Pipe package with spillway at outlet chamber. (*After ATV, 1978.*)

7.6 ACCESS OPENINGS

Normal cleaning and maintenance of a pipe package requires that access openings, (i.e., manholes) be provided. Such openings at the inlet and at the outlet chambers are required for

- entry of personnel to the chamber,
- transport of equipment and materials,
- ventilation, and
- light shafts.

When pipe packages have more than three parallel pipes, it is recommended that two openings be installed in each chamber.

Experience in Europe and in the United States reveals that normal cleaning equipment can service approximately 80 to 100 linear feet from the opening. As a result, pipe packages exceeding 200 feet in length need additional access openings along each of the pipes in the package.

7.7 OPERATION AND MAINTENANCE

Like any other storage facility, pipe packages need regular maintenance. To insure this, a formal maintenance schedule should be prepared which also provides for routine inspection. Each time the detention facility is visited, it should be inspected for

- clogged outlets and obstructed inlets,
- excessive deposits,
- corrosion of metal parts,
- deterioration of concrete, and
- any other damage or visible problems.

The outlet can experience momentary clogging by large objects such as lumber, cans, balls, plastic sheets, etc. This type of problem can be significantly reduced by providing at least two outlet openings at two different levels in the outlet chamber.

Clogging at the outlet can also be caused by buildup of sludge, sediments, and debris. Regular inspection and the removal of accumulated deposits and debris is the most effective way to deal with such problems.

REFERENCES

ATV, *Guidelines for Design and Configuration of Storm Water Relief Facilities in Mixed Water Channels,* Arbeitsblatt A 128, 2nd edition, 1978. (In German)

MLYNAREK, L., "Stormwater Impoundment Channels of Asbestos Cement Pipes," Korespondenz Abwasser No. 4, 1975. (In German)

RANDOLPH, R., BORDERS, J., AND HELM, R., "Pipe Storage for Equalization of Backwash Discharges to Sanitary Sewers," *Journal of Water Pollution Control Federation,* No. 8, 1980.

8

Tunnel Storage

A conveyance tunnel for storm or combined sewage must, for technical reasons, be designed much larger than required to handle the design flow. As a result, tunnels often have significant excess volume that can be used for storage. Since the inflow into a tunnel usually takes place through drill holes or drop shafts, it is also possible to use large surcharge depths in tunnel systems. This chapter describes various tunnel storage configurations and also provides a brief description of the technical configuration for each of the arrangements.

8.1 CLASSIFICATION BY FUNCTION

Tunnel storage can be classified into three functional categories:

- detention,
- off-line storage, and
- sedimentation.

8.1.1 Detention Tunnels

Detention tunnels are used frequently in Sweden to convey wastewater flows to treatment plants. A detention tunnel is a conveyance tunnel that also detains stormwater or wastewater behind some form of a flow restrictor.

Often, the flow restrictor is designed to pass the dry weather flows undetained. At the same time, it is the configuration and the size of the flow restrictor that defines how detention takes place. It is safe to assume that a detention tunnel will have a continuous dry weather flow, as can be seen in Figure 8.1, and this low flow has to be accounted for in the design of the outlet.

8.1.2 Off-line Storage Tunnels

Off-line storage tunnels are used in several cities in Europe and in the United States. An example of this is the case where an overloaded combined sewer is relieved by diverting excess flows to a tunnel that is off-line to the main flow in the sewer. This type of installation differs from the one described in section 8.1.1 in that it has no dry weather flow except for groundwater seepage.

Water stored in an off-line tunnel does not interact with the flow in the sewer network until it is routed back into the network. Typically, stored water is emptied from the tunnel using pumps after the flow in the sewer has subsided. A schematic representation of this type of an installation is shown in Figure 8.2.

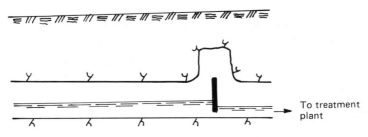

Figure 8.1   Detention tunnel.

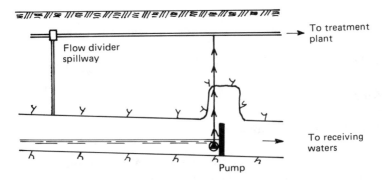

Figure 8.2   Off-line storage tunnel.

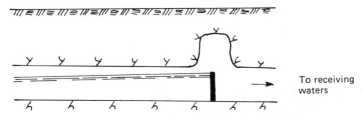

Figure 8.3   Tunnel as a sedimentation basin.

8.1.3 Sedimentation Tunnels

Tunnels can also be used as sedimentation basins. This is achieved by detaining the water in the tunnel so that settling of sediments can occur. The outlet of a sedimentation tunnel is usually equipped with a small orifice and an overflow weir so that all significant flows will be released over the weir. This type of an arrangement is illustrated in Figure 8.3.

8.2 DETENTION TUNNELS

Detention tunnels are in essence very low velocity conveyance conduits that provide detention by backing up water inside a large boring behind some type of a flow regulator. For the most part, detention tunnels are used to regulate or equalize storm flows in a combined sewer system.

8.2.1 Configuration of Flow Regulators

One common feature of flow regulators in detention tunnels is to permit dry weather flows to pass through unobstructed. Only when a desired flow rate is reached, the flow is held back by the regulator and detention occurs. The following are some examples of flow regulation in tunnels for which design details are discussed in Part 2 of this book:

1. *Restricted Outlet.* The most common flow regulator is a restricted outlet, such as an orifice, a nozzle, a gate, or a valve.
2. *Reduced Pipe.* A pipe having a smaller diameter than the tunnel.
3. *Pumps.* The most positive control can be achieved using only pumps to empty the tunnel.

8.2.2 Spillways

To avoid excessive surcharge depths, or to permit outflow when the flow restrictor clogs, a spillway or an emergency outlet has to be provided. The spilled water can either be returned to the tunnel downstream of the regulator or it can be routed to the receiving waters untreated. This decision will depend on local site conditions and tunnel configuration.

8.2.3 Example: Flow Regulator with Fixed Gates

A flow regulating structure with fixed gates was installed inside one branch of a combined sewage tunnel network in Gothenborg, Sweden. Flow restriction is provided by two rectangular manually controlled gates. Each 40-inch by 40-inch gate is installed in a concrete wall that is approximately two-thirds the height of the tunnel (see Figure 8.4).

The frequently occurring low flows are not impeded by these gates and detain only the larger flows. When the headwater behind the gates exceeds the height of the concrete wall, water spills over the wall and continues downstream. Sludge deposits upstream of the gates have been small and have not been a problem so far.

8.2.4 Example: Regulation with Moving Gates

The inflow into the Kappala sewage treatment plant, which is located outside of Stockholm, Sweden, is regulated with movable gates. These gates, illustrated in Figure 8.5, were installed in 1973 in a tunnel immediately upstream of the plant and equalize the hydraulic load to the plant during storms.

The opening is controlled automatically in accordance with a preset operating routine. This routine takes into account the maximum flow into the treatment plant and the expected duration of peak inflows into the tunnel. The latter is estimated during each storm. Flow regulation usually occurs for only short periods (e.g., less than one day). On the other hand, during snow melt flows can be detained for five to seven days.

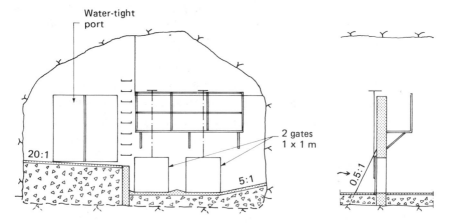

Figure 8.4   Flow regulator with fixed gates.

Figure 8.5   Flow regulating gate at the Kappala treatment plant.

Except for initial start-up problems, experience so far has been generally positive. One of the initial problems was the grit deposited in the tunnel overloading the plant's grit chambers when the gates were opened. This was solved by opening the gates more frequently to flush the grit accumulations.

During the first stage of testing, only 2.2 miles of tunnel was used for detention. At that time, the available detention volume was 530,000 cubic feet. Since 1980, up to 6.2 miles of tunnel has been used for detention, providing a volume of 1,400,000 cubic feet (i.e., 32 acre feet).

8.2.5 Example: Regulation with Pumps

At the Akeshov treatment plant in Stockholm, a pumping station with three centrifugal pumps lifts the water from a deep-lying tunnel system. The pumping height is approximately 100 feet, as shown in Figure 8.6. The tunnel is used as a detention facility to achieve a more uniform flow into the plant. Problems have been reported with the clogging of the suction lines of the pumps. This is understandable, since the water does not go through coarse separation of solids before pumping.

8.2.6 Example: Regulation with Reduced Pipes

A tunnel system that collects and transports combined drainage and wastewater in Gothenborg, Sweden crosses a major river. The river is crossed by a three-pipe inverted siphon. Each pipe is about 48 inches in diameter, and

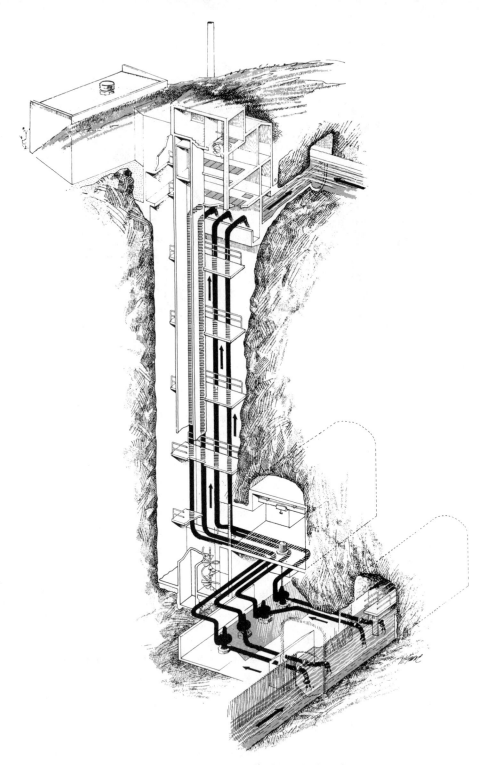

Figure 8.6 Pumping station at the Akeshov plant.

Figure 8.7   Tunnel-to-siphon transition in Gothenborg.

only two are connected at this time. The transition from the tunnel into a siphon occurs at a concrete wall, which is illustrated in Figure 8.7.

The siphon has significantly less flow capacity than the tunnel; thereby it acts as a flow restrictor. During high flows, the water backs up behind the siphon into the tunnel. An emergency spillway is located upstream of the siphon transition and provides relief when exceptionally high runoff occurs.

Initially, only one pipe in the siphon was connected to the tunnel. At that time, excessive sediment deposition was occurring in the tunnel. This deposition problem practically disappeared when the second siphon pipe was connected.

8.3 OFF-LINE STORAGE TUNNELS

Off-line storage refers to detention that occurs intermittently when a part of the flow in a collection system is diverted to a tunnel for temporary storage. Storage occurs only when the sewer system needs to be relieved. This means that, as a general rule, there is no base flow in these types of tunnels.

The water stored in an off-line storage tunnel has to be eventually returned to the sewer system and in most cases, this is done through the use of pumping. An exception is a system in very steep terrain where it could be drained by gravity.

The configuration of such a tunnel can vary from site to site. However, the following common elements can be found in many of the existing systems:

- a vertical drop shaft for the inflow;
- arrangements for emptying of the stored water;

- arrangements for ventilation; and
- provisions for the cleaning and removal of sediments.

8.3.1 Vertical Drop Shafts

Water is generally diverted to a tunnel through a number of points along the tunnel's length. These are customarily designed as vertical drop shafts. These shafts have to be designed to safely handle the maximum design inflow that may occur.

Occasionally, a vertical drop shaft can be designed to be used also for purposes other than inflow of water. These other uses can include descent openings for personnel, working pits, emergency evacuation points, air relief vents, etc.

Because of their deep and vertical configuration, the water flowing into the tunnel will have high kinetic energy at the bottom of the shaft. Because this high energy can cause structural damage to the tunnel and/or scour the rock out at the bottom of the shaft, some form of bottom reinforcement is needed at each drop shaft.

8.3.2 Draining the Tunnel

Because off-line tunnels tend to be located deep under the surface, they are usually drained through the use of pumps. The pumps and other appurtenances constitute the flow regulating structure, which can include:

- pumps for transferring water from the tunnel to the sewer system;
- a spillway to relieve excess inflow when full;
- pipes and pumps for decanting water from the tunnel to the receiving waters; and
- control equipment.

For a tunnel to act as a storage facility, the water has to be backed up by some form of a flow restriction at the downstream end. This can take a form of a concrete wall having a height to be determined by the following factors:

- slope of the tunnel;
- maximum allowable water level in the tunnel;
- magnitude of the inflow;
- capacity of the emptying facilities; and
- capacity of the spillway.

Tunnels used for temporary storage will also act as settling basins. As a result, if they store stormwater only, it is possible to decant the clarified water directly into receiving waters. Obviously, that is not the case with combined sewer systems.

Water stored in a temporary storage tunnel is usually transferred to the sewer network by pumping. Pump capacity and control is chosen in a way that the total system capacity is optimized. Following is a description of an optimizing procedure for the pumping of pollutants from an off-line tunnel to a treatment plant, as presented by Isgard in 1977.

Assuming that for a given pollutant the concentration C in a treatment plant's effluent, when treating combined sewage flows, can be expressed as a function of the inflow rate q into the plant, then

$$C = f(q) \tag{8.1}$$

Letting C_o be the effluent concentration when the plant effluent rate is q_o, the mass of the pollutant released during time T is

$$C_o \cdot q_o \cdot T \tag{8.2}$$

The volume of the stored water, V_s, to be pumped from the tunnel, is assumed to have concentration C_s of the pollutant, which corresponds to a mass of the pollutant equal to $C_s V_s$. If the stored water is pumped into the treatment plant over time T, the inflow into the plant will increase by

$$q_s = \frac{V_s}{T} \tag{8.3}$$

As a result, the total inflow into the plant during time T will be $(q_o + q_s)$ and the concentration of the pollutant in the effluent will increase to $(C_o + C_1)$. The mass of the pollutant leaving the plant over time T can be expressed as

$$(C_o + C_1) \cdot (q_o + q_s) \cdot T \tag{8.4}$$

In order to obtain any benefit at all from the storage (i.e., flow equalization), the following condition has to be met:

$$(C_o + C_1) \cdot (q_o + g_s) \cdot T \le (C_o \cdot q_o \cdot T + C_s V_s) \tag{8.5}$$

If the preceding condition is not satisfied, it can be argued that instead of storing the water in storage tunnel, it is better to release it untreated directly to the receiving waters. This is because the total mass of the pollutants entering the receiving waters is not reduced by treatment and storage. Equation 8.5 can be rewritten as follows:

$$C_1 \cdot (q_o + q_s) \le q_s \cdot (C_s - C_o) \tag{8.6}$$

Let

$$C_2 = (C_s - C_o) \tag{8.7}$$

then,

$$\frac{C_1}{C_2} \le \frac{q_s}{q_o + q_s} \tag{8.8}$$

If it can be shown empirically that

$$C_1 = C_o \cdot \left(\frac{q_s}{q_o}\right)^2 \tag{8.9}$$

and

$$C_s = K \cdot C_o \tag{8.10}$$

one obtains

$$q_s \leq q_o \, (\sqrt{K - 0.75} - 0.5) \tag{8.11}$$

By determining the analytic expressions for C_1 and C_s for different pollutants, it is at least theoretically possible to optimize the pumping rate from the storage facility. Since different pollutants will have different treatment rates, it is not possible to find a solution for Equation 8.11 that will work for all pollutants needing treatment. It should work, however, for the constituents of greatest concern.

When the inflow into a plant drops below a certain rate, the concentration of the pollutant in the effluent should remain practically constant. This is usually the dry weather design capacity of the plant, and it is best to utilize this capacity by emptying the storage during dry weather low flow periods. Unfortunately, this is not always practical, since storms can follow each other within short periods of time. As a result, storage facilities need to be emptied as soon as possible, taking into consideration the treatment plant's capacity to deal with the added hydraulic load.

8.3.3 Ventilation

When the tunnel is filling, air will be displaced corresponding to the volume of water entering the tunnel. In order to permit this air to be displaced and not be trapped, the tunnel has to be equipped with air vent openings. These openings have to be sized to accommodate the maximum permissible air velocity through the ventilation shaft, which normally should not exceed 33 feet per second. Since the tunnel usually fills faster than it empties, the vent is sized to accommodate the air flow during this phase of the operation.

In addition to providing vents for air relief, it is also necessary to provide air ventilation to all underground control and equipment rooms. These rooms need fresh air for the personnel and for the equipment that may be located within them. It is also recommended that these rooms be heated and well lighted to provide a safe and comfortable working environment.

8.3.4 Cleaning Off-line Tunnels

To keep the odors and corrosive gases under control, off-line tunnels have to be cleaned to remove sludge deposits that accumulate within them. If self-cleaning cannot be realized, the tunnel needs to be constructed with a hard bottom to accommodate mechanical cleaning equipment. Cleaning of tunnels can be very expensive; thus it is best to design the tunnels large enough to require infrequent cleaning (i.e., 5- to 15-year intervals). Also, due to the inherent hazard of the tunnel environment to humans, mechanical cleaning is preferred over manual cleaning. Section 8.4 describes the mass and composition of the sludge deposits that can be expected inside storage tunnels.

8.3.5 Example: Off-line CSO Storage Tunnel

An example of an off-line tunnel used for temporary detention of storm-water and combined sewer overflows is the Alvsjo-Malaren tunnel in Stockholm, Sweden. This installation is illustrated in Figure 8.8. It is 3 miles in length and extends from the fairgrounds in Alvsjo to Lake Malaren in the vicinity of a sewage treatment plant. The tunnel receives combined sewer overflows (CSO) from an area served by combined sewers and from another area served by a separate urban storm sewer system. Approximately 1,000 feet from the downstream end of the tunnel is the flow regulating control system, which is located entirely underground.

An emergency spillway structure backs up the water in the tunnel and keeps it from flowing into the lake. When the surcharge level in the tunnel is greater than the spillway, water begins to overflow into the lake. The spillway structure is also equipped with an outlet gate, which opens automatically

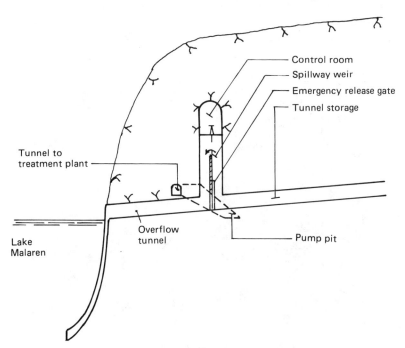

Figure 8.8   Alvsjo-Malaren tunnel. (*After Isgard, 1977.*)

during unusually large storms when the spillway overflow cannot handle the hydraulic load by itself. The decision when to open the auxiliary gates is based on how fast the water rises in the upstream end of the tunnel.

During typical wet weather operation, the tunnel is emptied into the sewer system for conveyance to the treatment plant. The tunnel can be drained to a certain level by gravity, and the rate of outflow is controlled by flow regulating valves. The remainder of the water has to be emptied by pumping. It is also possible to decant the cleaner surface water and to release it directly into Lake Malaren.

A picture of the control room for this installation is shown in Figure 8.9. Among other things, the picture shows the upper part of the emergency release gate. Note that all control equipment is enclosed in a small heated and ventilated building. The pressure pipes from the submersible pumps can be seen in Figure 8.10. They are also located close to the emergency outlet gate.

8.3.6 Example: Storage of Stormwater

After extensive drainage and water quality studies, it was decided that separate urban stormwater runoff from a new subdivision of Jarvafeltet in Stockholm, Sweden should be temporarily stored in a tunnel system. Because of the elevations, the tunnel had to be emptied by pumping. This installation, in addition to detention, was to act as a sedimentation tunnel.

The main body of the tunnel is about 4.3 miles long. It has seven tributary branches having a combined length of 2.5 miles, and four transport tunnels with a total length of 0.6 miles. In total, the system has 3.9 miles of

Figure 8.9   Alvsjo-Malaren tunnel control room.

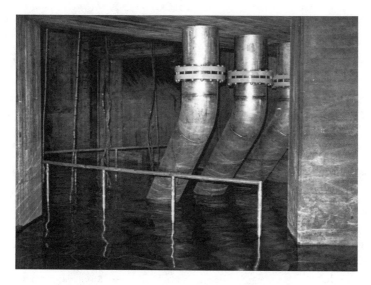

Figure 8.10   Alvsjo-Malaren tunnel pressure pipes.

tunnel with a cross-section area of 183 square feet, and 3.5 miles of tunnel with an area of 357 square feet. This corresponds to total volume of almost 10 million cubic feet (220 acre feet). Stormwater is conveyed to the tunnel system by 36 storm sewers or bore holes. Figure 8.11 shows a plan of the tunnel system.

The tunnel is emptied into the river Igelbacken through the Eggeby pumping station. The station has four pumps with a total capacity of 18 cubic feet per second plus one reserve pump. This pumping station is illustrated in Figure 8.12.

Because the tunnel will also function as a sedimentation basin, its bottom was paved with asphalt. This was done to prepare it for mechanical cleaning of deposited sediments and sludge. Operational experience to date indicates that it will have to be cleaned at 5- to 10-year intervals.

8.4 TUNNELS AS SEDIMENTATION BASINS

Tunnels used as sedimentation basins, like conventional sedimentation basins, are normally kept full of water in between storm events. When new water enters the tunnel, it displaces the water stored in the tunnel. The amount of cleansing the water receives is a function of the quality of the influent, the holding time in the tunnel, the velocity of flow in the tunnel, particle size distribution, and many other factors. Due to the oxygen-depleting nature of the deposits from combined sewage–stormwater and the difficulties associated

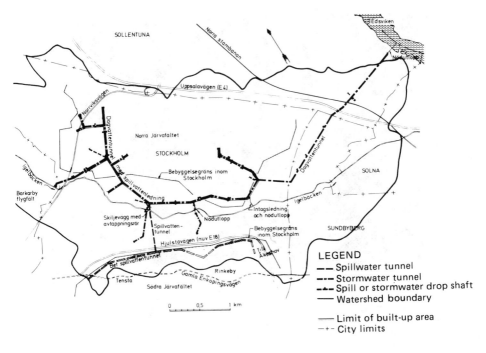

Figure 8.11   Tunnel plan for an area in Stockholm.

with these deposits, sedimentation tunnels should be used preferably for separate stormwater systems.

For the most part, sedimentation tunnels are self-regulating. Although the physical layout and arrangement of facilities may vary between sites, some of the common elements found in most sites include:

- drop shafts for stormwater inflow;
- ventilation and air relief shafts along its length;
- spillway weir that acts as the main flow regulator;
- pump(s) or gravity pipes for emptying of the tunnel for cleaning and maintenance; and
- tunnel floor suitable for mechanical cleaning.

8.4.1 Sedimentation

A yearlong study of the sedimentation efficiencies in a stormwater sedimentation tunnel was conducted in Sweden by the Stockholm Water and Sewage Works between September 1976 and September 1977. The water quantities entering and leaving the tunnel were recorded and water samples

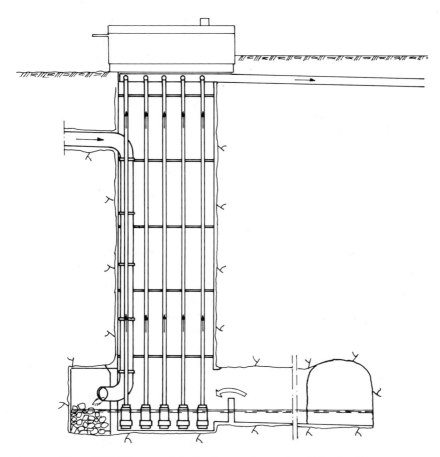

Figure 8.12 Diagram of the Eggeby pumping station. (After Stockholm Water and Sewage Works.)

were taken and analyzed. It was discovered that groundwater accounted for 25% of all the water entering the tunnel during the year. Since it was not feasible to readily measure the water quality of this groundwater, calculations of pollutant removal rates were made for the following two possible scenarios:

1. The pollutants in the groundwater were assumed to be zero.
2. The pollutant concentrations in the groundwater were assumed to be equal to those found in the stormwater.

The results of this study are summarized in Table 8.1. The low number in the range corresponds to the assumption that the pollutants in the groundwater are zero. The high numbers correspond to the assumption that groundwater

TABLE 8.1   Separation of Pollutants in a Settling Tunnel

	PERCENT REMOVAL	
Constituent	Tunnel	Laboratory
Dry residual solids	(−12)–13	11
Annealing solids	(−7)–18	—
Suspended solids	73–80	69–81
Ammonia as N	59–70	—
Nitrate and nitrite as N	0–23	—
Total nitrogen	6–29	7
Phosphate as P	55–64	—
Total phosphorus as P	50–63	48
Chemical Oxygen Demand (COD)	20–40	57–64

After Stockholm Water and Sewage Works, 1978

pollutant concentrations equal those found in the stormwater. This table also contains results of laboratory settling studies using the same stormwater. The laboratory studies used seven days as the sedimentation period, which corresponded to the average holding time in the tunnel.

The negative separations reported for the dry residual and the annealing solids resulted because of the assumption that the concentrations of pollutants in the groundwater were zero. The laboratory tests showed that the true reduction of these constituents was near the upper limit of the removal range measured in the tunnel.

The study also showed that stormwater was clarified for the most part during the first day of detention. When the concentration of the suspended solids decreased to 10 to 20 milligrams per liter in the water, further sedimentation became extremely slow. Similar results were observed in the laboratory tests of the stormwater samples. Very similar observations were also reported a few years later by Randall et. al. (1982) for stormwater detention pond studies in the United States.

In addition to the reported results from the aforementioned Swedish study, estimates were also made of removals for some of the other constituents found in stormwater. However, due to the limited data obtained during these tests, the conclusions should be treated as preliminary. Parallel laboratory tests of these additional constituents showed that it took four to seven days of detention to achieve the reported results. These tentative findings are summarized in Table 8.2.

8.4.2 Sludge Deposits

As part of the aforementioned investigation, the rate of sediment (sludge) deposition was also investigated. Based on calculated estimates, the layer of deposits inside the tunnel increased in depth between 6 and 13 millimeters per year (i.e., 0.25 to 0.50 inches per year). This study also predicted that deposition of bottom deposits will not be uniform throughout the tunnel and will vary according to the location in the tunnel.

Table 8.3 summarizes the chemical analysis of the deposits collected in two stormwater tunnels in Sweden. For comparison, the table also shows the analytical results of digested wastewater treatment plant sludge. It can be seen that stormwater deposits, except for lead, manganese, and cobalt have concentrations for all pollutants that are less than are found in digested wastewater treatment plant sludge.

TABLE 8.2   Separation of Metals in a Settling Tunnel Based on Limited Data

Constituent	PERCENT REMOVAL	
	Tunnel	Laboratory
Lead	76–82	62–98
Cadmium	74–82	—
Copper	16–36	57–82
Zinc	48–60	50–55
Thermostable coliforms	96–97	99
Fat and oil	97–98	—
BOD_5	58–68	32–73
Turbidity	34–65	42–98

After Stockholm Water and Sewage Works, 1978

TABLE 8.3   Analysis of Stormwater Deposits in Two Tunnels

Constituent	Jarva Tunnel	Sollentuna Tunnel	Digested Sludge
Dry substance—%	31	32	—
Fat and oil—mg/(kg DW)	14,000	—	—
Hydrocarbons—mg/(kg DW)	8,000	21,000	—
Arsenic—mg/(kg DW)	7	—	—
Lead—mg/(kg DW)	420	580	100–300
Iron—mg/(kg DW)	42,000	48,000	—
Cadmium—mg/(kg DW)	3.7	5	5–15
Cobalt—mg/(kg DW)	36	24	8–20
Copper—mg/(kg DW)	120	150	500–1,500
Chromium—mg/(kg DW)	49	63	50–200
Mercury—mg/(kg DW)	0.036	0.22	4–8
Manganese—mg/(kg DW)	1,300	750	200–500
Nickel—mg/(kg DW)	52	49	25–100
Zinc—mg/(kg DW)	1,100	890	1,000–3,000
Phosphorus—mg/(kg DW)	5.4	—	—

According to Stockholm Water and Sewage Works
Note: DW stands for dry weight.

8.4.3 Groundwater Lowering and Terrain Subsidence

It is not possible to prevent groundwater from seeping into tunnels. The rate of the seepage is a function of the groundwater depth, the extent of rock fractures, and the extent of the measures taken to seal off the seepage. Seepage into tunnels can lower the groundwater table and, as a result, cause general subsidence of the surface.

It is possible to control groundwater lowering by keeping the tunnel under a hydrostatic pressure. Obviously, this is not practical for detention tunnels. On the other hand, it is possible to do so with sedimentation tunnels by raising the outlet so that the tunnel is under constant internal pressure.

8.4.4 Example; Tunnel as a Sedimentation Basin

In 1970, a stormwater tunnel was built in Sollentuna, a community located north of Stockholm, Sweden. The tunnel is approximately 1 mile in length, has a 75-square-foot cross-sectional area, is sloped at 0.5%, and is not sealed against groundwater infiltration. Since there was a risk of subsidence in a clay layer in the region, it was decided to keep the tunnel constantly under water pressure.

At the downstream end of the tunnel is an outlet and a spillway designed to keep the tunnel under positive pressure. The system is illustrated in Figure 8.13. The outlet was equipped with a gate that can be opened to prevent the water from rising too high. However, the gate is only to be used in the event of an extremely heavy inflow into the tunnel.

This tunnel functions only as a sedimentation basin for stormwater runoff. As a result, it has to be cleaned out whenever the sediment accumulation in the tunnel becomes excessive. The tunnel has to be drained to be cleaned so that the settled sludge can be mechanically removed. The tunnel was emptied for cleaning in 1977 after six years of operation. A layer of sediments was found to be 20 inches deep at the upstream end and 10 inches deep at the downstream end. As a result, cleaning was not considered necessary at that time. The tunnel was emptied again in 1979. The increase in sediment deposits, since it was inspected last, was judged insignificant.

8.5 HYDRAULICS OF TUNNEL STORAGE

Tunnel storage involves some special and sometimes unique hydraulic operational problems. Many of these were studied by the Department of Hydraulics Engineering at the Royal Institute of Technology in Stockholm, Sweden and addressed the following:

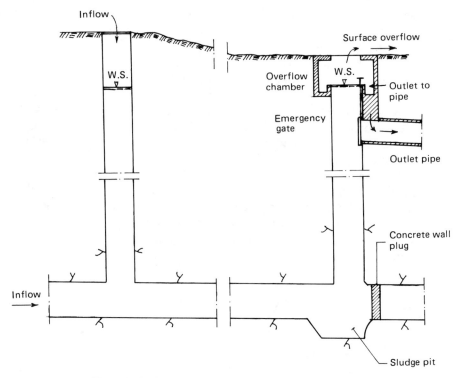

Figure 8.13   Sedimentation tunnel in Sollentuna, Sweden.

- hydraulic flow capacity,
- filling and emptying of tunnels,
- self-cleaning,
- air entrapment, and
- unsteady flow.

For a complete description of the study and its findings, refer to Bergh and Cederwall (1979).

8.5.1 Hydraulic Flow Capacity

The hydraulic flow capacity of a stormwater storage tunnel can be calculated using one of the empirical flow equations for steady uniform flow. The most widely used equation for this is Manning's formula:

$$Q = \frac{1.49}{n} AR^{\frac{2}{3}} S^{\frac{1}{2}} \tag{8.12}$$

in which Q = flow rate in cubic feet per second,
$\quad A$ = cross-section area of the tunnel in square feet,
$\quad R$ = hydraulics radius of the tunnel in feet,
$\quad S$ = longitudinal slope in feet per foot, and
$\quad n$ = Manning's roughness coefficient.

For circular sections, the flow capacity is greatest when the tunnel is 90% full. One should not design the tunnel at this theoretical maximum capacity depth. Instead, it is safer to design the tunnel to be at capacity when it is running full.

Stormwater storage tunnels are often designed to be lined, at least in part, with concrete and to have a V-shaped bottom similar to the one illustrated in Figure 8.14. Manning's coefficient n for this type of a section is a composite of the coefficient for rock and for concrete portions of the cross-section. The composite Manning's n can be estimated with the aid of the following formula:

$$ n = \left[\frac{\dfrac{n_c^{3/2}}{P_c} + \dfrac{n_r^{3/2}}{P_r}}{P_c + P_r} \right]^{\frac{2}{3}} \tag{8.13} $$

in which n = Manning's composite roughness coefficient,
$\quad n_c$ = Manning's coefficient for concrete,
$\quad n_r$ = Manning's coefficient for rock,
$\quad P_c$ = wetted perimeter for concrete, and
$\quad P_r$ = wetted perimeter for rock.

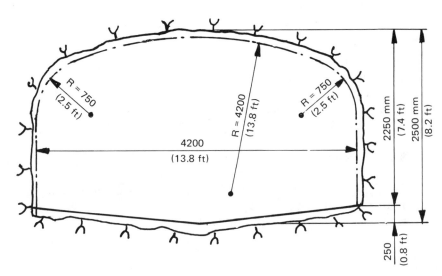

Figure 8.14 Typical stormwater storage tunnel section. (All measurements in millimeters.)

For tunnels having cross-section areas between 50 and 160 square feet, the Manning's *n* for rock can be set at 0.030. For the concrete liner in the tunnel, the Manning's *n* can be set at 0.013.

8.5.2 Filling and Emptying a Tunnel

The design of a storage tunnel has to account for the way a tunnel fills with stormwater and how it is emptied. If the filling and emptying process is not adequately accounted for, stormwater can surcharge the system or cause waters to back up into areas being protected from flooding.

As an example, the hydraulic response of a tunnel equipped with an emergency outlet gate, similar to the one illustrated in Figure 8.8, was simulated using a computer. The results of how the water surface reacted during a storm are illustrated in Figure 8.15, which shows the water level variations in the tunnel after the emergency gate is opened. The gate was opened when the water level at the upstream end of the tunnel rose to a level twelve meters above the outlet.

Although the water level in some parts of the tunnel dropped rather quickly, the level at the inlet continued to rise. This time delay in the response has to be recognized both in the design and in the operation of the tunnel. Therefore, extensive dynamic computer modeling of such facilities is recommended when preparing the final design.

8.5.3 Self-Cleansing

When water flows, a shear stress occurs between the water and any surface it touches. This is universally true, including for flow in tunnels. In the

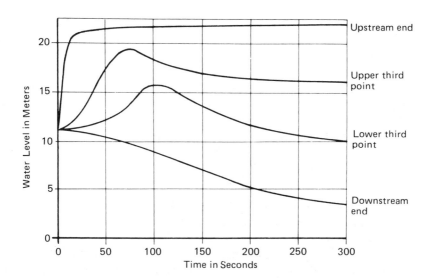

Figure 8.15   Water levels after downstream gate is opened.

case of uniform flow, the shear stress on the wetted perimeter of a tunnel can be calculated using the following formula:

$$T_{ave} = \rho \cdot g \cdot R \cdot S \qquad (8.14)$$

in which T_{ave} = average shear stress on the wetted perimeter,
   p = density of water,
   g = acceleration of gravity,
   R = hydraulic radius, and
   S = slope of the tunnel.

This shear force also acts on the materials that settled on the bottom of a tunnel. For these particles to be moved by the flow, a critical shear stress has to be exceeded. A tunnel can be considered self-cleansing if the average shear stress, T_{ave}, along the bottom of the tunnel is greater than the critical shear stress.

Lysne (1976), guided by the results obtained by a number of different researchers, proposed that the critical shear stress is $4 \ N/m^2$ (i.e. Newtons per square meter) for stormwater sewers and $2 \ N/m^2$ for sanitary sewers. The values are probably applicable in tunnels if self-cleansing of the tunnel is to be achieved.

8.5.4 Air Venting

Air that is trapped in a tunnel reduces both the available storage volume and the flow capacity of a tunnel. Also, compressed air can have a considerable amount of energy. For instance, 1,300 cubic yards of air compressed under 80 feet of water, if released suddenly, can produce the same explosive force as 100 pounds of dynamite.

Evacuation of compressed air that has been compressed inside of a tunnel can result in strong pressure waves in the tunnel and in the bore hole shaft (see Figure 8.16). As a result, water can surge out of the shaft in a form of a geyser. Such a phenomenon was observed at the Sollentuna tunnel north of Stockholm, Sweden. When this happened, a manhole lid weighing 110 pounds was thrown almost 3 feet into the air by a jet of water that eventually reached a height of 25 feet.

Air can enter the tunnel in the following ways:

- air entering with the stormwater;
- mixing of air into water at points of entry into the tunnel;
- mixing of air at hydraulic jumps inside the tunnel; and
- mixing of air in conjunction with pumping.

The best remedy is to prevent air from entering the tunnel, but this is virtually an impossible task. As a result, it is necessary to remove air from the tunnel by installing air traps at strategic locations inside the tunnel. Air can be trapped by installing vented enlarged sections and by venting it at other points inside the tunnel, such as at bends. The water can slow down at these locations and air can rise to the surface, thus escaping through the vent shafts (see Figure 8.17).

As mentioned earlier, rapid air evacuation can create shock waves. To avoid this, each evacuation shaft can be equipped with a sleeve that projects down into the tunnel having a series of air metering holes located near the ceiling of the tunnel. Air is thus released in a controlled fashion through the vent shaft (see Figure 8.18).

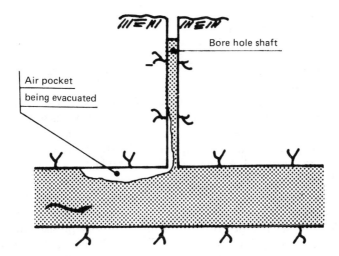

Figure 8.16   Air evacuation from a completely filled tunnel. (After Bergh and Cederwall, 1978.)

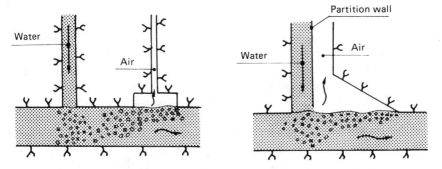

Figure 8.17   Evacuation of air through air traps. (After Bergh and Cederwall, 1978.)

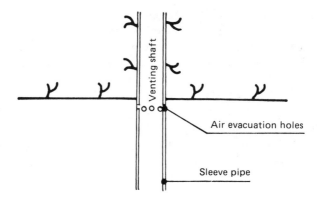

Figure 8.18   Air evacuation through venting in a sleeve pipe. (After Bergh and Cederwall, 1978.)

8.5.5 Unsteady Flow

As the tunnel fills or empties, it creates unsteady flow inside the tunnel. This is observed when gates are opened or closed, or when sudden inflow of stormwater occurs. As a result, the analysis of flow in tunnels cannot be performed using only steady flow equations. Unsteady flow is more properly described with partial differential equations, which can be solved using numerical methods.

Using computers to solve partial differential equations is a common practice these days. However, such numerical solutions can become unstable, and when these instabilities occur, the solutions can predict unreasonable water levels and flow fluctuations. We urge that the user always evaluate all numerical solutions for reasonableness before accepting the results.

REFERENCES

APWA, *Feasibility of Utility Tunnels in Urban Areas,* Special Report No. 39, 1971.

APWA, *Proceedings of the Conference on Engineering Utility Tunnels in Urban Areas,* Special Report No. 41, 1971.

BERGH, H., AND CEDERWALL, K., "Hydraulic Operating Problems in Storm Water Tunnels," KTH, Vattenbyggnad, Stockholm, 1978. (In Swedish)

FLOOD CONTROL COORDINATING COMMITTEE, *Development of a Flood and Pollution Control for Chicagoland Area,* Summary of Technical Reports, August, 1972.

ISGARD, E., "Undergrond Storage in Sewerage Systems," Rock Store, Proceedings of the First International Symposium, Vol. 1, Stockholm, 1977.

LYNAM, B., NEIL, F., AND DALTON, F., "Rock Tunnel Contracts Being Awarded for the Metropolitan Sanitary District of Greater Chicago Tunnel and Reservoir Plan (TARP)," Tunneling Technology Newsletter No. 17, March, 1977.

LYSNE, D. K., "Self-Cleaning in Runoff Pipes," PRA Report No. 9, 1976. (In Norwegian)

PARTHUM, "Building for the Future, The Boston Deep Tunnel Plan," *WPCF Journal,* April, 1970.

RANDALL, C. W., ELLIS, K., GRIZZARD, T. J., AND KNOCKE, W. R., "Urban Runoff Pollutant Removal by Sedimentation," *Proceedings of the Conference on Stormwater Detention Facilities,* American Society of Civil Engineers, New York, 1982.

STAHRE, P., "The Use of Tunnel Systems for Storage of Sewage Water," Rock Store, Proceedings of the International Symposium, Vol. 1, Stockholm, 1980.

STOCKHOLM WATER AND SEWAGE WORKS, Brochures, Drawings, etc.

STOCKHOLM WATER AND SEWAGE WORKS, *Studies of Stormwater Quality in the Jerva Drainage,* Internal Report, 1978. (In Swedish)

UNIVERSITY OF WISCONSIN, *Deep Tunnels in Hard Rock. A Solution to Combined Sewer Overflow and Flooding Problems,* College of Applied Science and Engineering, EPA Report 11020-02/71.

9

Storage at Sewage Treatment Plants

9.1 INTRODUCTION

The inflow into a sewage treatment plant can exhibit considerable variations over time. These variations can be significant in a smaller plant and can cause operational disturbances in the treatment process.

The variations in the flow can be balanced, to a certain degree, by installing some form of storage at the treatment plant. This storage can be in the form of:

- separate storage basins, or
- storage integrated into the plant itself.

9.2 EFFECTS OF FLOW VARIATION ON TREATMENT PROCESSES

9.2.1 General

The various treatment processes at a plant are not equally sensitive to the variations in the inflow. How sensitive each of the following treatment processes is to flow variations is discussed next:

- pretreatment,
- primary treatment,
- biological treatment, and
- chemical treatment.

9.2.2 Pretreatment

The pretreatment process consists of the screening out of coarse particles and the removal of sand and grit. None of the processes are especially sensitive to variations in flow or in the pollution load. Also, oversizing of these pretreatment facilities can be accomplished at a relatively small increment of added cost.

In new treatment plants, the pretreatment facilities should be installed ahead of any flow equalization storage. This assures reduced maintenance costs and means that the continuous operation of the basin is considerably simplified.

9.2.3 Primary Treatment

The primary treatment takes place in a treatment plant between the pretreatment and the biological or chemical treatment processes. Primary treatment consists of sedimentation and removal of the suspended solids in the wastewater. If this process is hydraulically overloaded, wastewater will pass through the primary tanks too fast for efficient settling to occur. As a result, the primary treatment process will not remove its full share of the pollutants and will pass them on to the subsequent treatment processes.

As with the pretreatment process, increasing the capacity of the primary treatment facilities is relatively inexpensive. Thus, it is good practice to provide sufficient primary treatment capacity to handle the anticipated maximum peak flows. Also, for some installations, flow equalization can be incorporated into the primary treatment tanks. This practice can provide a uniform stream of flow and pollutants to the downstream treatment processes.

9.2.4 Biological Treatment

Biological treatment usually occurs after the primary treatment process. It is normally accomplished in two steps. The first step transforms the organic matter carried by the waste water into microorganisms, while the second step removes these microorganisms from the water through settling. Because the growth of microorganisms is relatively slow, best treatment is obtained when the microorganisms are provided uniform living conditions (i.e., exposure to uniform flow and pollutant load).

Not all biological processes are affected equally by variations in flow. The sensitivity of treatment to inflow variations differs between the activated sludge process and the trickling filter and the rotating biological disk processes. The sensitivity of the biological process is said to depend, to a large extent, on the following factors:

1. the ability of microorganisms to break down organic matter;
2. the ability of microorganisms to form flock (i.e., sludge), which can readily settle out; and
3. the efficiency of the sludge settling unit.

In the activated sludge process, the ability of the organisms to break down organic matter is influenced by how much the rate of flow and pollutant load varies with time. Rapid variations cannot be accommodated efficiently. Variations in loading will also cause the development of sludge which will settle out very poorly.

Flow variations will also decrease the efficiency of the sludge settling units (i.e., clarifiers). As the flow rate reaches a critical level, a massive breaking off of the biological flock occurs and some, or all, of the sludge is flushed out of the clarifiers. If too much sludge is flushed out, the entire activated sludge process is disrupted for a long period of time.

In the trickling filter and the rotating biological disk processes, the microorganisms are normally under a heavier load than in the activated sludge process. Because the biological activity takes place on the surface of the filter or rotating disk media, variations in wastewater flow will have little effect on the ability of the flock to settle out in the clarifiers. With increasing flow, the efficiency of the biological process of the trickling filter and the rotating biological disk will decrease. There is, however, no critical flow level at which massive breaking off of the biological flocks occurs.

9.2.5 Chemical Treatment

The efficiency of chemical treatment, for the most part, is dependent on the following three factors:

1. precipitation of phosphates;
2. ability to form easily settlable flock; and
3. efficiency of the settling unit.

The efficiency of phosphate removal is only marginally affected by the flow rate or the quality of the wastewater. This assumes, however, that the chemical dosage is adjusted to the variation in the load. Nevertheless, the holding time in the flocculation basin is directly dependent on the inflow rate, and the ability to form flock is influenced by variations in flow.

The efficiency of the sedimentation basin to separate the chemical flock from the water is, however, directly affected by the flow rate through the basin. As with biological clarifiers, the vertical velocity of flow has to be less than the settling velocity of the flock for separation to occur. Generally, the chemical flock has greater settling velocities than the biological sludge, and the chemical settling basin can handle greater variations in flow. As the flow increases, the settling basin efficiency will decrease; however, even if massive loss of flock occurs, the total treatment process efficiency is not compromised for an extended period of time, as can happen in a biological process.

9.3 SEPARATE STORAGE BASINS

Flow equalization of inflow into a wastewater treatment plant can be accomplished with separate storage basins at the treatment plant. As illustrated in Figure 9.1, the basins can be connected either in-line or off-line to the main flow entering the plant. Regardless of the connection scheme, the storage basins are located between pretreatment and the primary treatment processes. This assures that most of the coarse floatables, sand, and grit are removed before the flow enters the storage basin. If, however, the basin has to be

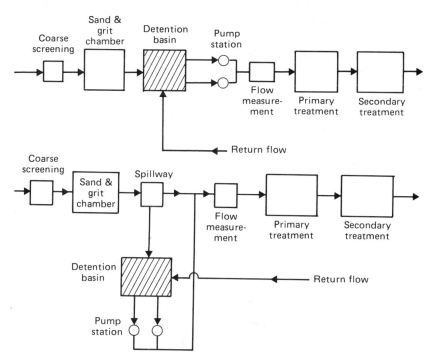

Figure 9.1   Flow equalization location in treatment plants. *(After EPA, 1974.)*

located ahead of the pretreatment unit, the basin needs to be equipped with equipment to remove sludge from its bottom.

9.3.1 In-line Basins

When the storage basin is connected in-line with the inflow, all the wastewater entering the treatment plant will flow through the basin. The flow out of the basin is then regulated through the use of either automated gates or pumps. The use of automated gates is only feasible when gravity flow into the subsequent treatment processes is possible.

Figure 9.2 shows a diagram of an in-line storage facility with a gravity driven outflow. The outflow is regulated by an automated gate valve that maintains a constant flow rate. Whenever the inflow rate exceeds the outflow rate, water is stored in the basin, and when the inflow rate is less than the outflow, the stored water is released. If the plant operators can predict the flow variations in advance, the outlet rate can be preset to balance out all of the flow peaks and valleys during the day.

Whenever unusually heavy flows occur, the storage volume will be exceeded. This will cause the basin to overflow, and the flow will continue downstream to the next treatment process. Because of the likelihood of this happening in a combined wastewater system, it is a good idea to provide an emergency overflow spillway.

For gravity flow to be feasible, the treatment plant site needs to have considerable vertical gradient. This is not possible at most sites, and supplemental pumping is required. Most plants utilize pumping as the primary means of controlling the flow rate into the treatment plant. Such an arrangement is illustrated in Figure 9.3. When the gradient through the plant is insufficient to permit gravity flow, it is necessary to provide emergency power and pump redundancy to ensure continuous operation of the system.

9.3.2 Off-line Storage Basins

Off-line storage basins are filled only when a preset rate of flow into the plant is exceeded (see Figure 9.4). The inflow to the storage facility is regulated by a specially designed side channel spillway. The water that is stored in

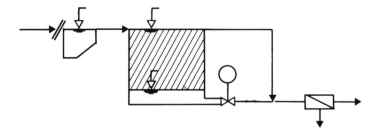

Figure 9.2 Storage in series with gravity outflow. *(After Nyseth, 1980.)*

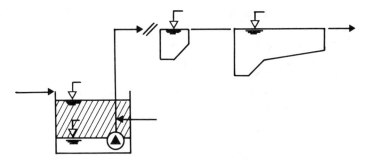

Figure 9.3   Storage in series with pumped outflow. *(After Nyseth, 1980.)*

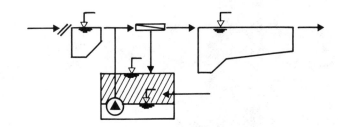

Figure 9.4   Storage in parallel to inflow to the plant. *(After Nyseth, 1980.)*

the basin is pumped to the main inflow stream when the inflow rate falls below a preset rate.

The off-line arrangement permits the dry weather wastewater to flow directly to the treatment processes. Since storage occurs only when flows spill into the basin, pumping requirements are less than for an in-line basin. Also, for the same reason, off-line basins are less prone to sludge accumulation than in-line basins.

9.3.3 Basin Configuration

The configuration of a separate storage basin is, for the most part, determined by site conditions. Among the factors affecting the configuration of a basin are

- storage volume needs,
- space availability, and
- vertical grade at the site.

Small- and medium-sized basins are often constructed out of structural concrete. Large basins, however, take the form of lined open ponds. Both were discussed earlier in Chapters 4 and 5.

A circular concrete detention tank is illustrated in Figure 9.5. It was under construction and was awaiting the installation of a rotating sludge scraper. In Figure 9.6, one can see an open concrete lined pond. The accumulated sludge in this pond is removed after the pond is drained using front end loaders and trucks.

For storage basins having long holding times, it may be necessary to install mixing and aerating equipment. This equipment may be needed to

Figure 9.5   A 3 acre-foot (4,000 m³) tank under construction.

Figure 9.6   A 10 acre-foot (13,000 m³) open storage pond.

prevent anaerobic conditions from developing inside the basin. See Metcalf and Eddy, Inc. (1979) or other wastewater treatment design texts for guidance in how to size such aerators.

9.4 INTEGRATED STORAGE

In some cases, it may be possible to incorporate the flow equalization storage into the basins used for primary treatment and aeration. This is feasible only when the variations in the flow are not very large. A detailed discussion on this topic can be found in Speece and La Grega (1976) and Spring (1977).

9.4.1 Flow Equalization in the Aeration Tanks

Flow equalization storage may be integrated into the aeration tanks of an activated sludge process. This is a possibility where the holding time in the aeration tanks is long and a relatively large portion of the basin volume can be used for detention. As Figure 9.7 illustrates, the water surface in the aeration tank will fluctuate with the flow.

Clearly, the biological processes will be affected by the variation in the water volume within the tank. To keep a stable biological system, the plant operators have to balance the hydraulic volume, pollutant load, the air supply, and the sludge mass being returned from the clarifiers. Careful monitoring of these parameters and of the average age of the sludge is necessary to maintain a stable biological treatment process, although it may be difficult with rapidly varying flow.

The main advantage of using aeration tanks for flow equalization is that the clarifier capacity is increased. However, as the sludge concentration entering the clarifiers increases, its hydraulic load has to be decreased to prevent solids from being flushed downstream. The main disadvantage of this approach is that the operation of the activated sludge process becomes significantly more complicated.

9.4.2 Flow Equalization in Primary Treatment Tanks

Flow equalization can also be provided in the primary treatment tanks of the treatment plant. This is accomplished, as illustrated in Figure 9.8, by constructing the tanks with excess freeboard. The flow is regulated by pumping the tanks at a relatively constant rate.

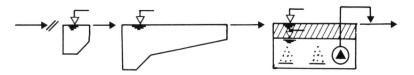

Figure 9.7   Flow equalization within an aeration tank.

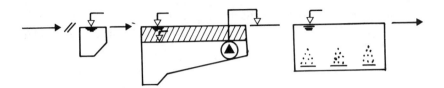

Figure 9.8   Flow equalization within a primary treatment tank.

The advantages of this concept include minimum land area requirements within a plant site and a relatively steady flow through the activated sludge process. It is possible, however, that the efficiency of the primary treatment process may be significantly reduced if very large inflow peaks are encountered.

REFERENCES

EPA, *Flow Equalization*, EPA Technology Transfer Seminar Publication, 1974.

FOESS, G., MEENAHAN, J., AND BLOUGH, D., "Evaluation of In-Line and Side-Line Flow Equalization Systems," *WPCF Journal*, January, 1977.

JUNKSGAARD, D., "Flow and Load Variations at Wastewater Treatment Plants," *WPCF Journal*, No. 8, 1980.

LA GREGA, M., AND KEENAN, J., "Effects of Equalizing Wastewater Flows," *WPCF Journal*, January, 1974.

METCALF & EDDY, INC., *Wastewater Engineering: Treatment Disposal and Reuse*, latest edition, McGraw-Hill.

NYSETH, I., "Use of Detention Basins at Sewage Treatment Plants," NTNF:s Utvalg for Drift av Renseanlegg, Prosjektrapport 23, 1980. (In Norwegian)

SPEECE, R., AND LA GREGA, M., "Flow Equalization by Use of Aeration Tank Volume," *WPCF Journal*, November, 1976.

SPRING, W., "Use of Free Capacities in the Pre-clarification by Recirculation of Storage," Kommunalwirtscaft H.9, 1977. (In German)

——— 10 ———
Other Types of Storage Facilities

10.1 INTRODUCTION

So far, only the more common types of storage facilities have been described. Naturally, there are other possibilities for storing stormwater or combined sewer overflows. Two examples of somewhat unconventional detention facilities are described here. In both cases, storage occurs within the receiving water body.

10.2 SUBMERGED CLOSED CONTAINERS

In the late 1960s, three full-scale tests were undertaken in the United States by the Environmental Protection Agency to collect and store combined sewer overflows in closed containers that are submerged in receiving waters. This technique was tested to see if such facilities could be installed in densely populated areas (see Figure 10.1). As a result of these tests, the following observations can be stated:

1. Submerged storage facilities are intended for the capture and storage of untreated combined sewer overflows that would otherwise enter the receiving waters.

Figure 10.1   Artist's rendering of a submerged storage. *(After Underwater Storage, Inc. and Silver Schwartz, Ltd., 1969.)*

2. These facilities are best suited in large, densely populated areas where open land near the receiving waters is not available.

3. This type of storage can be considered for areas such as harbors, industrial complexes, waterfront business centers, swimming beaches, promenades, etc.

4. There are no major technical problems in anchoring closed containers to the bottom of the receiving waters.

5. Because the detention container is entirely submerged, it has virtually no adverse aesthetic impacts on the landscape.

We describe here three configurations tested by U.S. Environmental Protection Agency for the collection and storage of combined sewer overflows. All three facilities were submerged storage containers fabricated, in part or entirely, out of flexible membrane materials.

10.2.1 Technical Arrangements

Storage of water in submerged closed containers will routinely require the following equipment:

- diversion chamber in the trunk sewer,
- flow measuring instrumentation,
- pretreatment of the water to be stored,
- a storage container,

- connecting pipes,
- pumps for emptying the container, and
- control equipment.

Figure 10.2 shows a diagram of how the various parts of the installation may be arranged. As can be seen, only the storage unit itself and the pipes are located in the receiving waters. All other equipment is contained in a building located on land.

When the flow in the trunk sewer has reached a predetermined rate, the excess flow is diverted to the submerged storage container. Before the flow enters the storage container, it first passes through a pretreatment chamber. Here the large particles are screened out and much of the silt and grit are removed from the flow. Just downstream of the pretreatment chamber, the pipes are equipped with valves to shut off flow to the storage unit. This is needed to permit the isolation of the unit for cleaning and repair.

Wastewater is stored in the storage container until the trunk sewer has sufficient capacity to accept additonal flow. The decision when to pump the stored water is made on the basis of information being transmitted by flow measuring equipment inside the sewers. The pumping equipment is located on land and not in the submerged storage units, which permits easy access for maintenance and repairs. The inlet pipe to the storage unit can also serve as the suction pipe for the pumps.

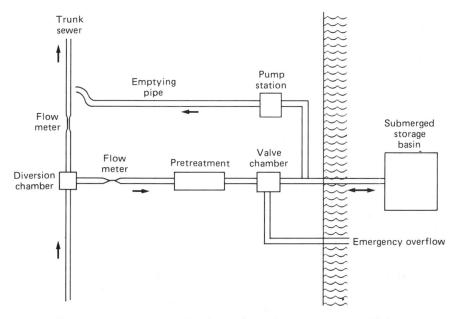

Figure 10.2 Arrangement of equipment for a submerged storage installation.

10.2.2 The Storage Unit

Table 10.1 summarizes some of the characteristics of the three EPA test installations built in the late 1960s. In all three, the storage units were fabricated, at least in part, using nylon-reinforced synthetic rubber membranes. As far as it is known, no other submerged storage basins have been constructed to date for the purpose of storing combined sewer overflows.

TABLE 10.1 Comparison of Three Submerged Storage Basins

	LOCATION OF THE INSTALLATION		
Item	Washington, D.C.	Cambridge, Md.	Sandusky, Ohio
Storage volume	Two 100,000-gallon units	One 200,000-gallon unit	Two 100,000-gallon units
Configuration	Rubber membrane anchored in steel cradle	Epoxy treated steel tank with rubber membrane top	Rubber membrane with steel frame inside and rigid bottom
Catchment area	30 acres	15 acres	20 acres
Distance to shoreline	115 feet	1,300 feet	On shoreline
Testing period	1969	1969	1968–1969

Despite pretreatment, suspended solids accumulate in these storage containers. Since the deposits can clog pipes and valves, it was found necessary to install some form of agitation in the storage containers. During the testing by EPA, the following were used to maintain the solids in suspension:

- circulation pumping of the stored water,
- blowing compressed air into the storage containers, and
- rinsing of the containers with high-pressure water.

At the installation illustrated in Figure 10.3, the storage container was flushed with water as it was emptied. To facilitate sludge removal, this storage unit was equipped with a pit at the inlet/outlet point within the unit.

Air and other gases that accumulate inside a storage container are evacuated using one-way air-relief valves installed at the top of each container. Since the container is submerged in a receiving water body, the valve permits the release of gases from the container while at the same time preventing the liquids from escaping. This valve also should not permit water from the outside to enter the storage container.

10.2.3 Operation and Maintenance Experience

A common feature of all three test facilities was that they were in operation for a relatively short period of time. Certain difficulties arose with continuous operation of all test units, which included:

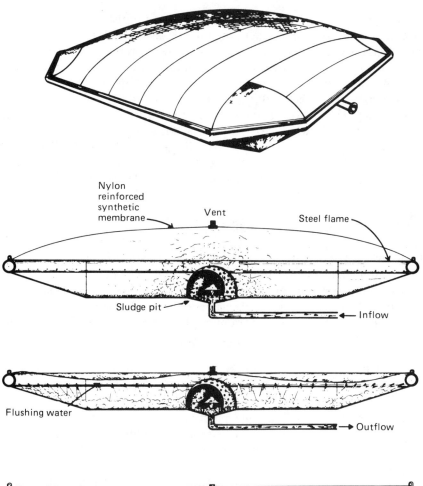

Figure 10.3   A sketch of the installation at Sandusky, Ohio. *(After Karl R. Rohrer, Inc., 1971.)*

- sediment deposits in the storage unit;
- damage to the rubber membrane from outside causes;
- failures of air-relief valves;
- anchoring of the storage containers to the bottom;
- clogging of the inlet/outlet pipes; and
- malfunctions of the land-based mechanical equipment.

It is only fair to mention, however, that most of these problems can be attributed to lack of experience at the time of testing with this type of technology. It is not unusual to experience "shakedown" problems whenever a new system goes on-line. It is likely that many of the aforementioned "problems" were solved since the tests were conducted in the late 1960s.

We discussed earlier in this book the importance of public perception in the design and installation of detention facilities. This point was driven home at one of the test sites when extremely strong negative public reaction forced a premature end of the test and the dismantling of the installation. The lesson learned from this experience is that it is important to consider more than purely technical issues when locating and designing detention facilities.

10.3 IN-LAKE FLOATING BASINS

In late 1970s, another unconventional storage/treatment facility was developed in Sweden to treat separate stormwater and combined sewer overflows. This technique was originally developed to reclaim heavily polluted lakes that had a limited number of annual exchanges of water. The storage unit consists of linear floats anchored to the bottom by plastic curtains, is always filled with water, and has an open connection to the lake.

10.3.1 Layout and Operation

The storage unit is made of floating docks that serve as the linear floats. The docks are held in place by steel pilings driven into the bottom of the lake and form a frame to which are attached plastic membrane curtains. The curtains are anchored to the bottom by concrete weights or chains, as illustrated in Figure 10.4.

The storage unit is divided into several flow-through basins that are connected in series. Water passes from one basin to the next through openings in the plastic curtain. An example plan of such a facility is shown in Figure 10.5.

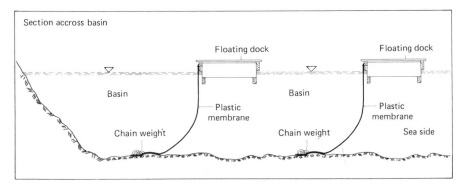

Figure 10.4 Cross-section of in-lake floating basins. *(By Soderlund, 1982.)*

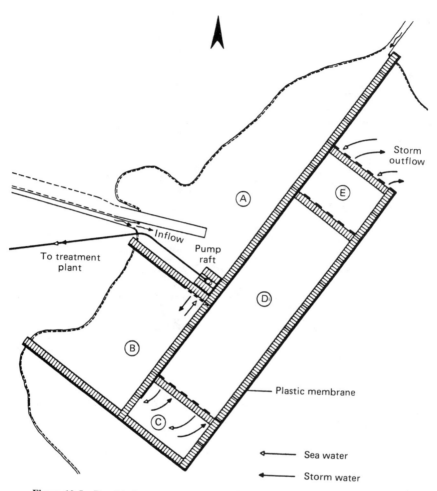

Figure 10.5 Possible layout of an in-lake floating basin. (After Soderlund, 1982).

When stormwater runoff is flowing into the lake, it first enters basin A, from which water is pumped continuously to a treatment plant. During very heavy runoff, the inflow exceeds pump capacity and runoff enters the next downstream basin, B. If heavy runoff continues, all cells, namely A through E, will be filled with runoff and will enter the lake.

As the runoff passes through each storage cell, it displaces the lake water. While displacing lake water, some mixing of the lake's water with the runoff occurs. When storm runoff subsides and the inflow drops below the pump capacity, water then flows back into cell A, where it is pumped to the treatment plant. During dry weather, the treatment continues and lake water is drawn into the treatment plant. This process was designed to remove the phosphorus from the lake and help arrest the eutrophication process.

10.3.2 Field Experience

Through 1980, three in-lake storage facilities were built near Stockholm, Sweden. Their characteristics are summarized in Table 10.2. As part of the evaluation of this technology, flow mixing patterns were studied in detail at the basin installed in Lake Trehoningen. A complete report of the findings was prepared by Anderson (1980). Some of his key findings are summarized briefly here.

TABLE 10.2 Study Findings of Three In-lake Storage Units

	LOCATION OF THE INSTALLATION (LAKE)		
Item	Trehorningen Huddinge	Ronningesjon Taby	Flaten Stockholm
Basin area	29,000 ft^2	15,000 ft^2	60,000 ft^2
Volume	175,000 ft^3	81,000 ft^3	530,000 ft^3
No. of cells	15	5	2
Treatment	Chemical precipitation	Chemical precipitation and partial filtration	—
Capacity	8,500 ft^3/hr	2,500 ft^3/hr	—
Year installed	1978	1980	1980

Using radioactive tracers, leakage was found to occur between separation curtains. The degree of leakage depended on the location along the curtains separating adjacent basins. When leakage occurred, some of the inflow short-circuited the treatment process.

Anderson (1980) reported that the most common points of leakage were at the corners, where two or more curtains joined. This was attributed, in part, to the pressure gradients found between "kitty-corner" storage cells. It was observed that the curtains bulged in the direction of flow and caused gaps to open between the curtains at the corners. Cementing and better sealing at the joints should help reduce this type of short circuiting.

During very high runoff, Anderson (1980) observed that the pressure differential between cells was sufficiently large to lift the curtains off the bottom. When that happened, some of the water flowed under the curtains and short-circuited the storage facility. After the curtains had been lifted off the bottom, they were found to catch on the anchors and would not drop down to seal against the bottom. This resulted in a recommendation to modify the anchoring system.

Anderson (1980) also reported that wind can also set up disturbances in the storage cells. A relatively moderate wind was found to set the water into rotational motion within the installation. However, it is possible to arrange the cells in a way that minimizes the effects of such wind disturbances.

REFERENCES

ANDERSON, L., "Detention Basins in Lake Trehorning-Huddinge Municipality, Study of Water Exchange," KTH Avdelningen for Vattenvardsteknik, Stockholm, 1980. (In Swedish)

KARL R. ROHRER ASSOCIATES, INC., *Underwater Storage of Combined Sewer Overflows,* USEPA Report No. 11022 ECVO9/71, NTIS No. PB 208 346, September, 1971.

MELPAR: AN AMERICAN STANDARD COMPANY, *Combined Sewer Temporary Underwater Storage Facility,* USEPA Report No. 11022 DPP 10/70, NTIS No. PB 197 669, October, 1970.

SODERLUND, H., *Flow Balancing Methods for Stormwater and Combined Sewer Overflows,* Swedish Council for Building Research, Publication D17, 1982. (In Swedish)

UNDERWATER STORAGE INC., AND SILVER, SCHWARTZ, LTD., *Control of Pollution by Underwater Storage,* USEPA Report No. 11020 DWF12/69, NTIS No. PB 191 217.

11

Overview of Flow Regulation

11.1 INTRODUCTION

Detention of stormwater or combined wastewater–stormwater flows is provided by a storage volume that is released by some type of a flow regulating device. It is the flow regulator that determines how efficiently the storage volume is being utilized. Obviously, the flow regulator has to be in balance with the available storage volume for the range of runoff events it was designed to control.

The outlet structure or device, what we call here the *flow regulating structure,* is an important component of the total detention storage facility. It not only controls the release rate, but determines the maximum storage depth and volume at the detention site. The flow regulating structure often is called on to perform what may appear to be conflicting tasks, such as limit the flow rates, be free of clogging, be relatively maintenance free, be designed to provide safety to the public and, in some cases, be aesthetically appealing.

In Part 2 we deal primarily with the hydraulic function of flow regulation, which can be accomplished in a number of different ways. The chapters of Part 2 describe some of the conventional and some of the less conventional ways to provide flow regulation. As an introduction, some of the fundamental principles behind flow regulation are first examined.

11.2 FLOW REQUIREMENTS

The primary purpose for detention is to reduce the peak flow of runoff and equalize the rate of flow downstream. A rough estimate of the required storage volume can be made by assuming that the outflow rate is constant. However, a constant outflow rate cannot be easily achieved and most hydraulic structures have flow rates that vary with the depth of water. Obviously, the required storage volume for flow equalization is determined, to a large extent, by the outflow characteristics and the inflow hydrograph, as illustrated in Figure 11.1.

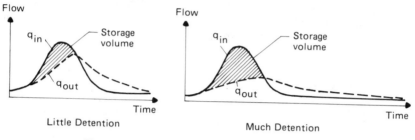

Figure 11.1 Storage volume needs vs. rate of outflow.

The specific configuration and details of an outlet structure will vary from site to site. However, for the more typical installations, the design of the operational outlet will try to do the following:

- Provide a maximum release flow rate for a single or multiple levels of design storm events.
- Minimize the storage volume by providing outlets to reach the maximum release rate early in the storm (see Figure 11.2).
- Permit flow-through of dry weather flows without backing up into the storage basin.
- Provide adequate trash racks to prevent clogging of the outlet.
- Where possible, provide the installation of simple, reliable, and operator-free equipment for regulating outflow.
- Provide good maintenance and inspection access, even when there is water in the storage basin.
- Allow for safety of the public and the maintenance and operating personnel.

11.3 LOCATION OF THE FLOW REGULATOR

The location of the flow regulating structure will depend on how the storage basin is connected to the storm conveyance system. If the storage is connected in-line (i.e., in series), the flow regulation takes place at the downstream end of the storage facility (see Figure 11.3). In this type of an arrangement, the flow regulator often is a flow restrictor located at, or near, the bottom of the storage basin. In some cases, where the topography does not permit emptying of the storage basin by gravity, pumping is used to regulate the flow rate.

When the storage facility is connected off-line (i.e., parallel) to the conveyance system, flow regulation is accomplished by limiting the rate of flow that bypasses the storage basin. The excess flow is diverted to the storage facility. Figure 11.4 shows two examples of how this system can be arranged. Other examples of how detention storage can be connected to the stormwater conveyance system can be found in Chapter 5 Section 2.

In practice, the diversion to storage in a parallel connected system occurs through a side-channel spillway. The bypass flow is regulated by some form of a hydraulic constriction just downstream of the diversion, as illustrated in

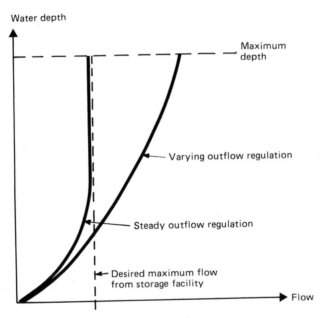

Figure 11.2   Example of steady outflow and variable outflow regulation.

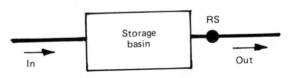

Figure 11.3   Location of a flow regulating structure in an in-line connected storage.

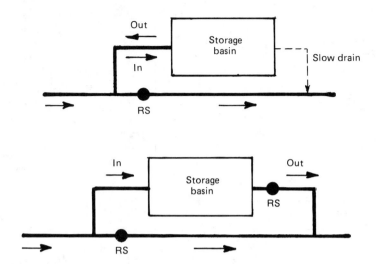

Figure 11.4 Location of a flow regulating structure in an off-line connected storage system.

Figure 11.4. Off-line storage normally requires a separate outlet to totally drain the impoundment, as depicted by the dashed line on the upper illustration in Figure 11.4. In some cases, the site topography will require the water to be pumped out of the storage basin. It may also be necessary to provide a separate flow regulating structure (i.e. RS on Figure 11.4) at the outlet of the storage basin.

11.4 FIXED AND MOVABLE FLOW REGULATORS

The release rate from a storage facility can be controlled with either a fixed or a movable flow regulating structure. Fixed regulators are permanently attached to the basin structure and have a constant outlet cross-section during the entire detention cycle. Fixed regulators include orifices, nozzles, weirs, and special, sometimes patented, devices. Except for some of the special flow regulating devices, the flow rate through a fixed regulator will vary with the hydraulic head.

Movable regulators are characterized by an outlet that varies in size or elevation during the detention cycle. This family of regulating devices include gates, valves, floating orifices, floating weirs, and pumps. All of these devices require some form of control equipment. Movable flow regulators can be self-regulating or they can be designed for remote control. Figure 11.5 depicts some of the methods used to regulate flows.

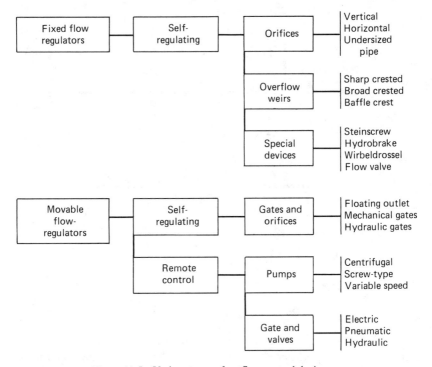

Figure 11.5 Various types of outflow control devices.

REFERENCES

LAGER, J. A., AND SMITH, W. G., "Urban Stormwater Management and Technology: An Assessment," U.S. EPA Report No. EPA-670/2-74-040, (NTIS No. PB 240 687).

══════ 12 ══════
Types of Flow Regulators

Flow regulators can range from a simple fixed orifice to a complex, remotely controlled, automated mechanically operated gate or valve. Although the discussion in this chapter describes many of the conventional and some of the less conventional flow regulators, it is not intended to describe every type of regulator that is available or that can be used.

12.1 FIXED REGULATORS

As stated earlier, fixed regulators have a flow control section that does not change in area or elevation during the detention cycle. Once constructed, modification of the control requires physical modification to the outlet. No special control equipment is needed, and these type of regulators are relatively inexpensive to build and operate. As was listed in Figure 11.5, fixed regulators include weirs, orifices, choked pipes, and special devices.

12.1.1 Vertically Arranged Orifice or Nozzle

The simplest flow regulating structure is an orifice or a nozzle installed into the side of a storage basin. The trickle flows flow through the opening unimpeded, while the larger flows are backed up (see Figure 12.1). When the outlet is small in comparison to the depth of water, the discharge through the orifice, or a nozzle, can be calculated using the following formula:

$$Q = C_d \cdot A \sqrt{2g \cdot (h - a)} \qquad (12.1)$$

in which Q = discharge rate through the outlet,
$\quad C_d$ = discharge coefficient,
$\quad A$ = area of the orifice or nozzle,
$\quad g$ = acceleration of gravity,
$\quad h$ = water depth at outlet, and
$\quad a$ = one-half the height of the outlet opening (see Figure 12.1).

The preceding equation assumes that there is no back pressure from downstream (i.e., the outlet is not submerged). If the outlet is submerged, this equation can still be used. Just use the difference in the water surfaces between both sides of the orifice as the depth h in the equation.

The discharge coefficient, C_d, can vary significantly with the shape and the type of the orifice. A summary of the most commonly used orifice shapes is presented in Figure 12.2. As a result, it is important to pay careful attention to the details of the orifice shape during construction to ensure that the completed facility will operate as designed.

It is sometimes necessary to estimate the time it takes to drain a known

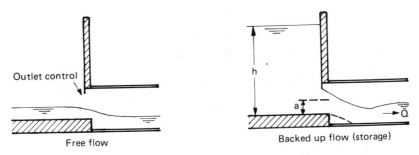

Figure 12.1 Outflow through a vertical orifice.

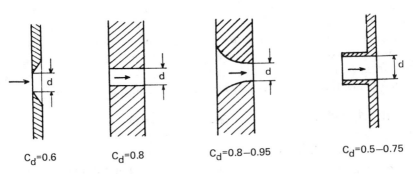

Figure 12.2 Discharge coefficient C_d for different outlets.

stored volume through an orifice. This is to ensure that the pond is emptied either sufficiently fast or slow after the storm ends. We start this calculation with the fact that the time needed to drain down an increment of depth dh is equal to the volume $A_R dh$ divided by the discharge Q. Integrating this between water depths h_1 and h_2, we get:

$$t = \int_o^t dt = -\int_{h_1}^{h_2} \left(\frac{A_R}{Q}\right) dh \tag{12.2}$$

Substituting Equation 12.1 for Q yields a relationship for calculating the emptying time of the storage volume,

$$t = -\frac{1}{C_d A \sqrt{2g}} \int_{h_1}^{h_2} \left(\frac{A_R}{\sqrt{h}}\right) dh \tag{12.3}$$

When the area of the reservoir, A_R, can be described as a mathematical function of the water depth, h, then this equation can be solved. For the special case where the surface area is a constant (i.e., vertical walls), Equation 12.3 is reduced to

$$t = \frac{2A_R}{C_d \cdot A \cdot \sqrt{2g}} \cdot (\sqrt{h_1} - \sqrt{h_2}) \tag{12.4}$$

The opening of the outlet can be round, square, rectangular, or any other convenient shape. It can be a preset gate or a plate attached to a headwall located in front of the outlet pipe. The orifice in the latter case is cut into a plate which is then installed so that it can be easily removed for maintenance or replacement.

Outlets with openings of less than 6 inches are very susceptible to clogging. The chances of clogging can be significantly reduced by installing trash racks having a net opening that is more than 20 times the opening of the outlet. Larger outlets also need trash racks, but the ratio of the net area of the rack to the outlet area does not need to be as large.

When the size of the orifice is relatively large when compared to the water depth, orifice equations (e.g., Equations 12.1 through 12.4) are not accurate. When water depth above the invert of the orifice is less than two to three times the height of the orifice, more accurate results are possible using inlet control nomographs for culverts published by the Federal Highway Administration (1963) and by using the improved culvert inlet design procedures, which are also published by the Federal Highway Administration (1972). Nomographs for calculating the discharge of circular and rectangular culverts operating under inlet control conditions are reproduced here as Figures 12.3 and 12.4.

12.1.2 Horizontally Arranged Orifice

In some cases, the flow regulating orifice has to be installed in the bottom of the storage basin similar to a bathtub drain (see Figure 12.5). When the

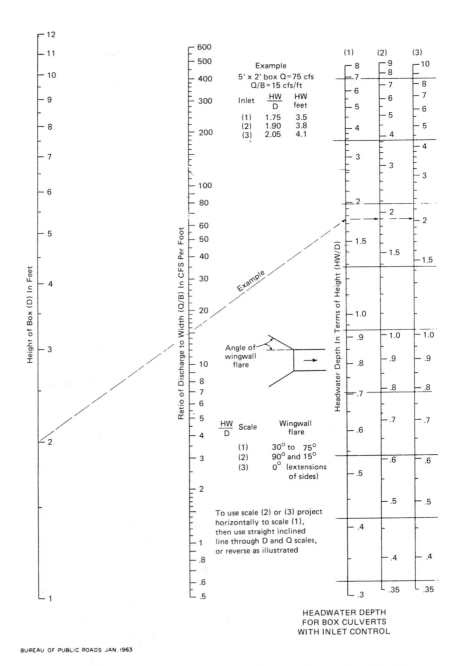

Figure 12.3 Nomograph for rectangular culvert capacity operating under inlet control. *(After Federal Highway Administration, 1974.)*

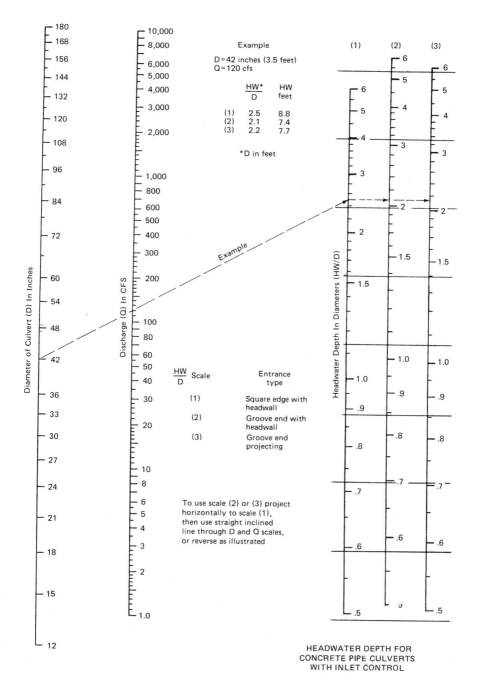

Figure 12.4 Nomograph for circular culvert capacity operating under inlet control. *(After Federal Highway Administration, 1974.)*

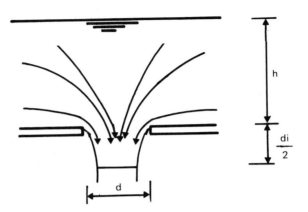

Figure 12.5   Flow through a horizontal orifice.

depth, h, is relatively large when compared to the orifice opening, its flow capacity can be calculated using the formula:

$$Q = C_d \cdot A \cdot \sqrt{2g \cdot \left(h - \frac{d}{2}\right)} \qquad (12.5)$$

in which Q = discharge rate through the outlet,
$\quad C_d$ = discharge coefficient,
$\quad A$ = area of the orifice or nozzle,
$\quad g$ = acceleration of gravity,
$\quad h$ = water depth at outlet, and
$\quad d$ = diameter of the outlet opening.

The discharge coefficients for a horizontally arranged orifice are the same as for the vertically arranged orifice, and representative values for both types can be found in Figure 12.2. Equation 12.5 will not give accurate results whenever the water depth above the horizontal orifice is less than three times the diameters of the orifice. At smaller depths, severe vortex action develops which is not accounted for in Equation 12.5.

12.1.3 Flow Restricting Pipe

There are at least two reasons why a flow restricting pipe may be used as an outlet. One is that it is difficult to modify the hydraulic capacity of an outlet pipe, unlike a flow restricting orifice which can be easily removed. Flow control orifices being removed by owners, or as an act of vandalism, has been reported by Prommersberger (1984) and others. The hydraulic characteristics of a flow restricting outlet pipe are much more difficult to modify. As illustrated in Figure 12.6, the net flow restricting effect of the pipe is mostly a function of the pipe length and pipe roughness characteristics.

A pipe outlet may also be used to provide greater flow reduction while

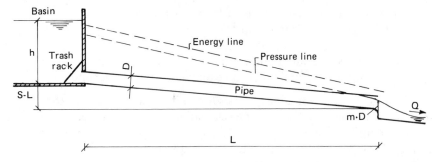

Figure 12.6   Flow regulation with an outlet pipe.

using a larger diameter outlet. If the pipe is set at a slope that is less than the hydraulic friction slope, outlet capacity can be reduced without the use of a small diameter orifice. It is important, however, to maintain a minimum velocity of 2- to 3-feet per second in the pipe in order to keep the silt carried by the water from settling out within the pipe.

If we assume that the pipe is flowing full and the discharge end of the pipe is not submerged, the outlet capacity can be calculated using basic hydraulic principles. If we begin with the continuity equation:

$$Q = A \cdot v \tag{12.6}$$

and calculate losses at the outlet using

$$H_t = K_L \cdot \frac{v^2}{2g} \tag{12.7}$$

combining these two equations yields:

$$Q = A \cdot \sqrt{2g \cdot \frac{H_t}{K_L}} \tag{12.8}$$

in which Q = outlet capacity,
   v = flow velocity when pipe is full,
   A = area of outlet pipe,
   g = acceleration of gravity,
   H_t = total hydraulic losses in a pipe outlet, and
   K_L = sum of loss factors for the outlet.

From Figure 12.6 we observe that,

$$H_t = h + S \cdot L - m \cdot D \tag{12.9}$$

Thus, Equation 12.9 can be substituted into Equation 12.8 and rewritten as

$$Q = A \cdot \sqrt{2g \cdot \frac{h + S \cdot L - m \cdot D}{K_L}} \tag{12.10}$$

in which h = depth of water above outlet pipe's invert,
  D = diameter of outlet pipe,
  S = slope of outlet pipe,
  L = length of outlet pipe, and
  m = ratio of water depth to pipe diameter at the outlet end of the pipe.

How to calculate outlet capacities for medium sized dams is explained in much detail by the U.S. Bureau of Reclamation (1973). However, some of the most commonly used procedures are repeated here. The sum of the loss factors will depend on the characteristics of the outlet. For example, it may contain

$$K_L = k_t + k_e + k_f + k_b + k_o \qquad (12.11)$$

in which k_t = trash rack loss factor,
  k_e = entrance loss factor,
  k_f = friction loss factor,
  k_b = bend loss factor, and
  k_o = outlet loss factor.

Trash Rack Loss Factor.   According to Creager and Justin (1950), the loss factor at a trash rack can be approximated using the following equation:

$$k_t = 1.45 - 0.45\left(\frac{a_n}{a_g}\right) - \left(\frac{a_n}{a_g}\right)^2 \qquad (12.12)$$

in which a_n = net open area between the rack bars, and
  a_g = gross area of the rack and supports.

When estimating the maximum potential losses at the rack, assume that 50% of the rack area is blocked. Also calculate the maximum outlet capacity assuming no blockage. Always calculate the minimum and maximum outlet capacities to ensure that the installation will function adequately under both possible operating scenarios in the field.

Entrance Loss Factor.   By taking the orifice equation, rearranging its terms, and recognizing that the depth term in the equation is actually the sum of the velocity head and the head loss, we find that

$$k_e = \frac{1}{C_d^2} - 1 \qquad (12.13)$$

in which C_d = orifice discharge coefficient.

Friction Loss Factor.   The pipe friction loss factor for a pipe flowing full is expressed as

$$k_f = f\frac{L}{D} \qquad (12.14)$$

in which f = Darcy-Weisbach friction loss coefficient, which, under certain simplifying assumptions, can be expressed as a function of Manning's n, namely,

$$f = 185 \frac{n^2}{D^{1/3}} \tag{12.15}$$

Bend Loss Factor.   Bend losses in a closed conduit are a function of bend radius, pipe diameter, and the deflection angle at the bend. The bend loss factor was studied by a number of investigators. Unfortunately, it was found to vary between different studies. However, for 90 degree bends having a radius at least twice the pipe diameter, a value of $k_b = 0.2$ appears to be a reasonable average using data from all of the studies. For bends having other than 90 degree bends, the bend loss factor can be calculated using the following equation:

$$k_b = K \cdot k_{90} \tag{12.16}$$

in which K = coefficient taken from Table 12.1, and
    k_{90} = loss factor for 90 degree bend.

TABLE 12.1   Factors for Other
than 90° Bend Losses

Angle of Bend in Degrees	Adjustment Factor
00	0.00
20	0.37
40	0.63
60	0.82
80	0.90
100	1.05
120	1.13

After U.S. Bureau of Reclamation, 1973

Outlet Loss Factor.   Virtually no recovery of velocity head occurs where the pipe freely discharges into the atmosphere or is submerged under water. As a result, unless a specially shaped flared outlet is provided, it is safe to assume that $k_o = 1.0$.

In calculating the pipe outlet capacity, it is generally assumed that the pressure line at the outlet end of the pipe is at the center of the pipe (i.e., $m = 0.5$). In certain cases, this assumption may not be valid. Li and Patterson (1956) developed Figure 12.7 to help estimate a correct value of m. First, find pipe capacity using equation 12.10, assuming that $m = 05$. Then find a new value of m from Figure 12.7 using the calculated velocity. Recalculate the pipe outlet capacity using equation 12.10 with the new value of m. One recalculation is sufficient to obtain an accurate answer.

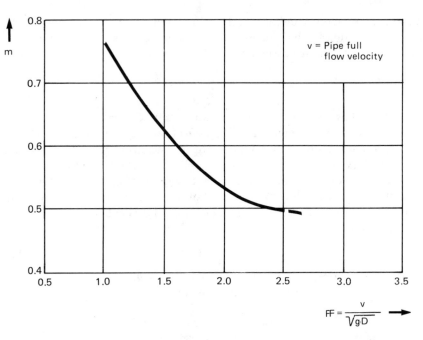

Figure 12.7 Froude Number vs. m at the pipe outlet. *(After Li and Patterson, 1956.)*

As stated, the outlet capacity is calculated assuming the pipe is flowing full. Figure 12.8 was also developed by Li and Patterson (1956) to help determine if the pipe is, in fact, entirely full. Although this figure is based on model tests using plastic pipe, it should provide reasonable basis for checking the flow condition in pipes manufactured of different materials.

12.1.4 Weirs

Weirs are often used as primary overflow control devices. They can provide the first level of emergency overflow, or actually be a part of the outflow regulating system. Weirs can come in many different shapes and sizes, and it is relatively easy to calculate the capacity of the more standard weir types. On the other hand, complicated compound weir configurations often do not have laboratory tested coefficients of discharge. As a result, calculations of their capacity may not be accurate. Fortunately, experimental data for many of the less standard shapes have been published by King and Brater (1963) and others.

Sharp-crested Weir. The flow over a sharp-crested weir having no end contractions can be calculated using the following equation:

$$Q = C \cdot L \cdot h^{3/2} \tag{12.17}$$

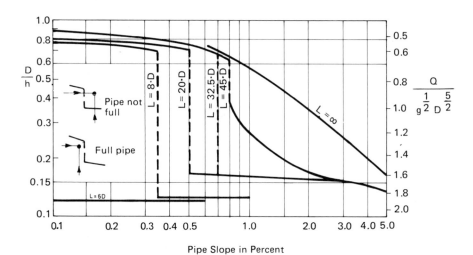

Figure 12.8   Length upstream of outlet needed to assure full pipe flow. *(After Li and Patterson, 1956.)*

in which Q = discharge over the weir,
   C = discharge coefficient,
   L = effective length of the weir crest (see Equation 12.19), and
   h = head above the weir crest.

The measurement of a representative head h for a sharp-crested weir is made at a distance of approximately $2.5 \cdot h$ upstream of the weir crest. Also, according to Chow (1959), the discharge coefficient for a sharp-crested weir can be calculated using

$$C = 3.27 - 0.4 \frac{h}{P} \qquad (12.18)$$

in which P = height of weir crest above the channel bottom (see Figure 12.9).

This equation gives accurate results if the weir nape is fully aerated and is not submerged. If the nape is not aerated, a partial vacuum develops under the nape and the flow over the weir increases. The flow also becomes very unstable and undulating. Such a condition is not desirable; therefore provisions for aerating the weir have to be made.

In most detention applications, the weir crest does not extend completely across the channel. Therefore, the length needs to be corrected for flow contractions at each end of a sharp-crested weir. The effective weir length is calculated using

$$L = L' - 0.1 \, n \cdot h \qquad (12.19)$$

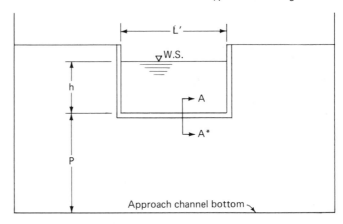

*See figure 12.10 for Section A-A

Figure 12.9   Sharp-crested weir.

in which L = effective length of the weir crest,
   L' = measured length of the weir crest,
   n = number of end contractions, and
   h = head of water above the crest.

Triangular Sharp-crested Weir.   A triangular sharp-crested weir should be considered whenever the weir needs also to control low flows. As can be seen in Figure 12.10, the water surface crest over this weir varies with depth. As a result, weir capacity is sensitive to the water depth at low flows. The discharge over a triangular sharp-crested weir is given by

$$Q = C_t \cdot h^{5/2} \cdot \tan\left(\frac{\alpha}{2}\right) \tag{12.20}$$

in which C_t = discharge coefficient for a triangular weir,
   α = weir notch angle in degrees, and
   h = head above weir notch bottom, in feet.

The head h is measured from the bottom of the notch to the water surface elevation at a distance of $2.5h$ upstream of the weir. Table 12.2 lists values of the discharge coefficient for the triangular weir and is based on the work reported by Lenz (1943) for a free, nonsubmerged nape downstream of the weir.

Submergence of a Sharp-crested Weir.   So far, we have described sharp-crested weirs that had a free, nonsubmerged nape on the downstream side. When the tailwater rises to above the weir crest (see Figure 12.11), the discharge calculations for the nonsubmerged case have to be corrected for sub-

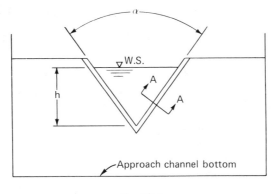

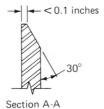

Elevation

Section A-A

Figure 12.10   Triangular sharp-crested weir.

TABLE 12.2   Discharge Coefficients for a Triangular Sharp-crested Weir

Depth h in Feet	COEFFICIENT C_t FOR NOTCH ANGLE			
	20°	45°	60°	90°
0.2	2.81	2.66	2.62	2.57
0.4	2.68	2.57	2.53	2.51
0.6	2.62	2.53	2.51	2.49
0.8	2.60	2.52	2.50	2.48

After Lenz, 1943

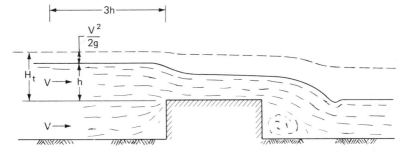

Figure 12.11   Broad-crested Weir.

mergence. Villemonte (1947) suggested the following equation for this purpose:

$$\frac{Q_s}{Q} = \left[1.0 - \left(\frac{h_s}{h} \right)^n \right]^{0.385} \tag{12.21}$$

in which Q = discharge calculated for a nonsubmerged weir equation,

$\quad Q_s$ = discharge for a submerged weir,

$\quad h$ = head upstream of the weir,

$\quad h_s$ = tailwater depth above the weir crest, and

$\quad n$ = the exponent in the sharp-crested weir equation (i.e., $\frac{3}{2}$ for rectangular and $\frac{5}{2}$ for triangular).

Theoretically, the preceding equation can be used to correct for submergence of any shape of a sharp-crested weir. All that is needed is to use the appropriate value for the exponent n. According to Villemonte, this equation has an accuracy of 5%.

Broad-crested Weirs. Broad-crested weirs are commonly used in stormwater storage facilities as overflow devices or spillways. The discharge over a broad-crested weir is given by the equation

$$Q = C \cdot L \cdot H_t^{3/2} \tag{12.22}$$

in which Q = discharge, ft³/sec,

$\quad C$ = coefficient of discharge,

$\quad L$ = effective length of the weir crest, ft,

$\quad H_t = (h + V^2/2g)$, total head above the weir crest, ft,

$\quad V$ = approach velocity at $3 \cdot h$ upstream of crest, ft/sec (usually taken at $V = 0$ for detention overflow), and

$\quad g$ = acceleration of gravity, 32.2 ft/sec.

The coefficient C for a broad-crested weir has been determined experimentally to range between 2.67 to 3.05. A value of $C = 3.0$ is often used for the design of detention overflow structures and spillways.

12.1.5 Special Flow Regulators

During the 1970s, a number of flow regulating devices were introduced with the intent of providing flow equalization. The most prominent of these are:

- Steinscrew,
- hydrobrake,
- wirbeldrossel, and
- flow valve.

All of these were designed to provide a more constant rate of outflow with varying water depth than is possible with an orifice or a flow restricting pipe outlet. Because of their unique configurations and designs, these special flow regulating devices are described in detail in Chapter 13.

12.2 MOVABLE FLOW REGULATORS

Movable flow regulators are classified into:

- self-regulating, and
- remote controlled.

Self-regulating flow regulators have an opening controlled by the water level in the storage basin or immediately downstream of the storage basin. The control is generally accomplished by the use of floats. No additional outside energy is required except for what is produced by the variations in the water level.

Remote controlled regulators are much more complex. Besides the equipment at the outlet itself, some form of electrical or pneumatic control equipment is needed. Its purpose is to give signals to the outlet structure to increase or decrease the release rate. Although remote controlled flow regulation can be expensive, it can be very effective in balancing the outflow rate against available sewer capacity downstream.

12.2.1 Flow Regulating Strategies

It is possible to "control" the release rate from the storage basin using only the water level upstream of the outlet. In this case, the flow downstream of the outlet is not monitored to see if the desired flow rate is being achieved. This approach can result in erroneous control signals and improper flow releases downstream.

As an example, if an outlet is partially clogged or silted in, the water level will rise in the storage basin. The control equipment interprets this water rise as an increasing release rate at the outlet. To keep the release rate constant, the control equipment throttles the outlet, further closing off the outlet. Obviously, the correct action should have been to open the outlet and flush out the clogging materials, or if it remains clogged, to signal the operator that there is a problem.

A much more effective flow control is possible by sensing the flow or water level downstream of the outlet. This permits direct interpretation of what is actually being released from the storage facility. By comparing the measured release rate with the desired flow rate, it is then possible to control the outlet to ensure that the downstream system capacity is being properly utilized.

In summary, for reasons of operational reliability, the release rate from a storage basin is best regulated using flow data obtained downstream of the outlet. One should avoid controlling movable flow regulators using only the water level inside of a storage basin.

12.2.2 Movable Gate with Mechanical Level Control

One type of movable self-regulating gate is controlled by the water level immediately downstream of the outlet. It consists of a stilling basin immediately downstream of the outlet and a float controlled flap gate. When the water level (i.e., the release rate) drops, the weighted float drops and opens the gate. When the water rises, the opposite occurs. This sequence can be set to provide a relatively constant release rate from a storage basin.

Experience has shown that operation can be impaired by the accumulation of sand, grit, and sludge in the float chamber. The valve eventually can be blocked from opening fully and the release rate diminished. This type of an installation requires considerable maintenance unless it is used in a sediment-free water.

12.2.3 Floating Outlet

A floating outlet is another type of a self-regulating movable flow regulator. The release from the storage facility occurs over an outlet that rises and falls with the water level in the storage basin. This permits the water depth at the outlet itself to be relatively constant, resulting in a uniform release rate.

An example of a floating outlet manufactured in England is illustrated in Figure 12.12. This outlet is equipped with a surface skimmer that keeps the floatables from being flushed downstream or from clogging the outlet itself. Since the water is skimmed off the top of the storage basin, much of the suspended solid load is trapped inside the basin. The device illustrated in this figure is manufactured with outlet pipes ranging between 3- to 15-inches in diameter and capacities of up to 2,500 gallons per minute (i.e., 5.6 cubic feet per second).

12.2.4 Remote Controlled Gates

Remote controlled gates vary from site to site, and their specific details depend on the conditions found at each location. With the advent of microprocessors and recent improvements in telemetering, the last 10 years have seen very significant advances in the technology of flow regulation. It is now

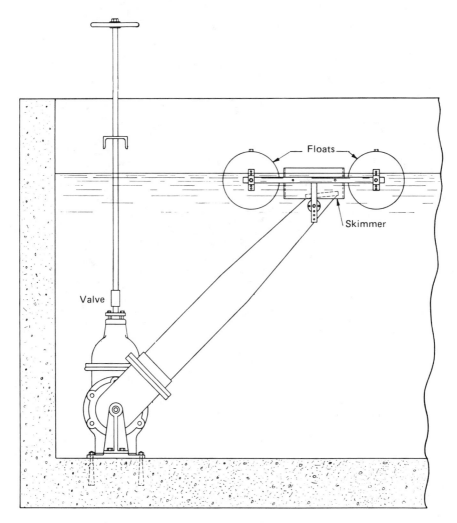

Figure 12.12   A floating outlet with a shut-off valve. *(Adams-Hydraulics, Ltd., York, England.)*

feasible to simultaneously control the release rates at a number of storage facilities for the purpose of optimizing total system capacity.

The basic principles for remotely controlled flow regulators are briefly described here. The reader is referred to papers by Shilling (1986) and others to learn more about the full potential of modern technology for this purpose. As shown in Figure 12.13, a remote controlled flow regulation system will typically consist of

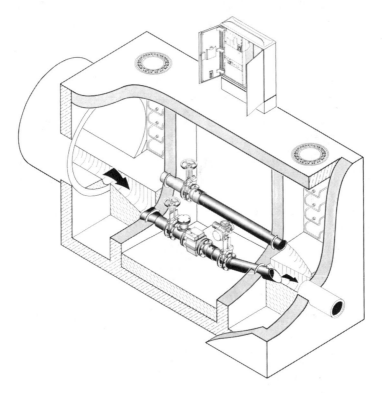

Figure 12.13 Flow measuring downstream of a storage basin. *(Umvelt und Fluidtehnik GmbH, Bad Mergentheim, West Germany.)*

- flow or water level sensing equipment,
- adjustable flow regulating gates, and
- a data processing unit.

   Flow Measurement.   The release from a storage basin is measured at a specially designed measuring section downstream of the outlet. Figure 12.13 shows one example of how flow sensing downstream on the storage basin can be arranged.

   Adjustable Gates.   Gates can be operated using electrical, hydraulic, or pneumatic controls. To improve operational reliability, gates should be installed in a separate access chamber located downstream of the storage basin. The automatic flow control gate is typically installed in a flow-through pipe located inside the vault. To permit the removal of the control gate for maintenance, it is recommended that a bypass pipe, equipped with a manually operated valve, be also installed (see Figure 12.13). Also, at least a 10-inch diameter gate is suggested to reduce clogging at the outlet.

Experience in Europe and in the United States (Schilling, 1986) clearly dictates that the control gate access vault be kept dry and dehumidified. Control and mechanical equipment deteriorate much more rapidly when the vault is humid and is subjected to inundation by wastewater. Good access, which is kept dry and free of humidity, is a cost-effective investment when one considers the consequences of equipment failures.

Control Processor.   Flow measurements or water levels at the monitoring section are transmitted to the processor unit. Here, data are transformed into signals for setting the opening of the control gate. When the measured rate exceeds the desired flow rate the gate opening is reduced, and when the flow rate is less than desired the gate is opened more. This form of regulation is referred to as a "feedback system."

Figure 12.14 illustrates the operation of a feedback system. The water level at a section downstream of the gate was used to control the opening of the control gate. When the water at the control section rises above h_{max} the processor tells the gate to close. The gate opening is reduced in small increments until the water level drops below h_{max}. Similarly, when the water level drops below h_{min} the gate is told to open more. Whenever the water level is between h_{max} and h_{min}, the gate opening is not changed.

During a large storm the release rate from the storage facility can rise very rapidly and exceed level $h_{max-max}$ at the control section. When this happens, the processor instructs the gate to close rapidly and brings the surcharge under control. Typically, the control system is designed not to fully close the

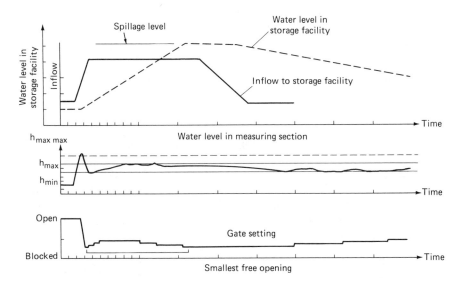

Figure 12.14   Example of feedback regulation of flow releases. *(After Jedelhauser, 1978.)*

gate. This minimum gate opening should be set to permit a desired release rate when the basin is full.

An automatic flow regulating system can be as simple as the one just described, or it can be made very sophisticated through the use of microprocessors, computer modeling, and telemetering. Such a system can be designed to "anticipate" the flow rate entering the storage basin and the release rate at the outlet by also providing flow measurements at the basin inlet, or by measuring the rainfall in the watershed. The latter requires simulation of the runoff process and is less reliable than inflow measurements. On the other hand, if the system needs more lead time to anticipate the control action to take, such as a storage basin in an undersized sewer network, the use of rainfall, runoff simulation, and flow verification may have to be an integral part of the control processor.

12.2.5 Flow Regulation With Pumps

Whenever the site topography does not permit draining of the storage basin by gravity, use of pumps is the only remaining option. The control processor is then used to operate pumps instead of gates. Because of the electrical and mechanical nature of pumps, the potential for electrical and mechanical failures is ever present and should be anticipated in design. Problems may include pump failure, electric power interruptions, clogging, and processor failures. Such problems can be handled by providing redundancy of equipment and by providing emergency power generation. With off-line storage facilities, operational reliability of pumps is further reduced by the intermittent nature of water storage in the basin.

The most common types of pumps used in flow regulation are:

- centrifugal pumps,
- eccentric screw pumps,
- screw pumps, and
- variable speed pumps.

Centrifugal Pumps. Centrifugal pumps generally are an excellent choice for the emptying of either in-line and off-line storage basins because of their reliability and relatively low maintenance needs. A uniform flow rate can be achieved by providing a depressed wet well which can maintain a relatively stable submergence on the pump impellers. Also, the outflow can be kept constant by a flow regulator that "throttles." The latter will result in the pump operating at less than its maximum efficiency.

Eccentric Screw Pumps. Continuously operating eccentric screw pumps will maintain a constant discharge rate, regardless of the water level in the storage basin. The disadvantage of this type of a pump is its relatively high cost and that the rotor is susceptible to wear. Due to their design, eccentric screw pumps have a tendency to clog more easily than centrifugal pumps.

Screw Pumps.   Screw pumps will maintain a constant discharge as long as there is enough water in the storage basin. In other words, screw pumps are hydraulically self-regulating. Their capacity is set by the slope of the screw trough, the rate of rotation, and the size of the screws. Although they are quite expensive, screw pumps offer operational advantages over other types of pumps used for this purpose. They are practically free from clogging, require a minimum of control equipment, and are usually very reliable to operate.

Variable Speed Pumps.   These pumps offer probably the greatest potential for flow control. The pump speed can be varied to increase or decrease the discharge rate to whatever the control processor demands. Because of their sophistication, variable speed pumps can be quite expensive to install and maintain. These pumps can be operated in combination with fixed speed pumps to maintain uniform flow.

REFERENCES

ADAMS-HYDRAULICS LTD, York, Great Britain, Product Brochures.

CHOW, VEN TE, "Open Channel Flow," McGraw-Hill, New York, 1959.

CREAGER, W. P., AND JUSTIN, J. D., *Hydroelectric Handbook,* 2nd edition, John Wiley and Sons, Inc., New York, 1950.

EPA, *Urban Stormwater Management Technology: An Assessment,* Environmental Protection Agency, May, 1979.

FEDERAL HIGHWAY ADMINISTRATION, "Hydraulic Charts for the Selection of Highway Culverts," *Hydraulic Engineering Circular No. 5,* U.S. Department of Transportation, Washington, D.C., reprinted 1974.

FEDERAL HIGHWAY ADMINISTRATION, "Hydraulic Design of Improved Inlets for Culverts," *Hydraulic Engineering Circular No. 13,* U.S. Department of Transportation, Washington, D.C., 1972.

FAHRNER, H., "Planning of Rain Basin Discharge Controls Using Automatic Throttle Discs," Abwasser No. 5, 1979. (In German)

JEDELHAUSER, H., "Outflow Control for Storm Water Settling Basins by Means of Pneumatic Slide Valves," Osterreichische Abwasser-Rundschau No. 2, 1978. (In German)

KING, H. W., AND BRATER, E. F., "Handbook of Hydraulics," McGraw-Hill, New York, 1963.

KORAL, J., AND SAATCI, A. C., "Storm Overflow and Detention Basins," 2nd edition, Zurich, 1976. (In German)

LENZ, A. T., "Viscosity and Surface Tension Effects on V-notch Weirs," *Transactions of the American Society of Civil Engineers,* vol. 69, New York, 1943.

LI, W. H., AND PATTERSON, C., "Free Outlets and Self-Priming Action of Culverts," *Journal of Hydraulics Division,* ASCE, HY 3, 1956.

PROMMERSBERGER, B., "Implementation of Stormwater Detention Policies in the

Denver Metro Area," *Flood Hazard News,* Urban Drainage and Flood Control District, Denver, Colo., December, 1984.

SCHILLING, W., "Urban Runoff Quality Management by Real-Time Control," *Urban Runoff Pollution,* NATO ASI Series G: Ecological Sciences, Vol. 10, Springer-Verlag, Berlin, 1986.

U.S. BUREAU OF RECLAMATION, *Design of Small Dams,* pp. 469–85, United States Government Printing Office, Washington, D.C., 1973.

VILLEMONTE, J. R., "Submerged-weir Discharge Studies," *Engineering News-Record,* p. 866, December, 1947.

————— 13 —————
Special Flow Regulators

In recent years, we have seen the development and marketing of flow regulators designed to provide a more uniform release rate under a varying depth. Here we describe four special regulators that have received considerable attention. These are not the only ones worth considering, but they are the ones we are most familiar with. Each of the following is a self-regulating unit requiring no mechanical control:

- Steinscrew,
- hydrobrake,
- wirbeldrossel, and
- flow valve.

13.1 STEINSCREW

The Steinscrew was developed in Sweden in the 1970s. It was designed to regulate wastewater flow rates in combined sewer pipes. The detention, as originally intended, occurs in the pipe itself, and no additional storage facility is needed. As a result, it is recommended that the Steinscrew be installed in large pipes (i.e., 36 inches or larger) to provide the flow balancing volumes.

There is no upper limit on the pipe size for which it would work, and it could be used in storage tunnels.

13.1.1 Technical Configuration

Steinscrew consists of a screw-shaped plate that is twisted into a 270° spiral and attached to the pipe as illustrated in Figure 13.1. The device occupies 90 percent of the pipe's diameter with a gap left between its longitudinal side and the crown of the pipe. The other longitudinal side is attached with stainless steel bolts to the bottom of the pipe.

The screw-shaped plate is equipped with two obliquely directed wings and form four triangular overflow sections between the plates and the wings, see Figure 13.2. In the bottom of the screw there is a circular opening to pass base flows. This base flow opening is made sufficiently large to permit self cleansing of the pipe's invert. Typically, the length of the Steinscrew is approximately three times the diameter of the pipe.

The Steinscrew is manufactured from a steel plate which is carefully shaped into the desired form. After welding and grinding, it is galvanized and epoxy-coated to protect it against corrosion. This treatment has proved to be sufficient protection in most installations.

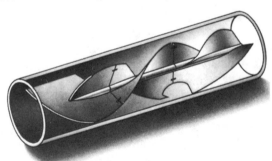

Figure 13.1   "Steinscrew" flow regulator. (After Janson and Bendixen, 1975)

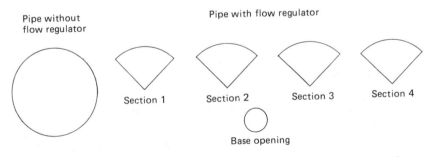

Pipe without flow regulator

Pipe with flow regulator

Section 1   Section 2   Section 3   Section 4

Base opening

Figure 13.2 Flow conveyance sections of the "Steinscrew". (After Janson and Bendixen, 1975)

13.1.2 Function

Whenever the flow exceeds the capacity of the low flow opening, water backs up into the pipe or into a tunnel. The amount of detention that takes place depends on pipe diameter and slope. Figure 13.3 can assist in estimating how much detention volume is available upstream of the Steinscrew when the pipe is 75% full just upstream of the regulator. It is recommended that the system be designed to operate so as not to exceed the 75% depth during the design rainstorm. More depth of storage in the pipe will cause the system to surcharge frequently, which is an unacceptable design condition.

As long as the water depth in the pipe is less than one-half pipe diameter,

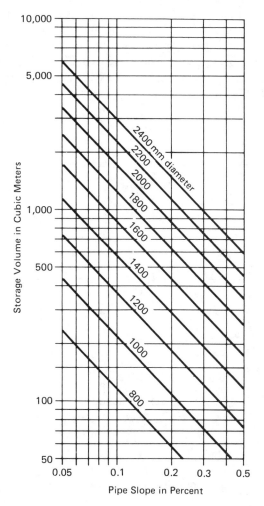

Figure 13.3   Storage volume in a pipe upstream of a Steinscrew.

all of the flow passes through the base opening. When the water rises further, it starts to spill through the triangular overflow sections, as illustrated in Figure 13.2. As the water continues to rise, the flow conveyance area rapidly increases until the flow capacity of the Steinscrew is almost equal to the capacity of the unobstructed pipe. See Figure 13.4 for the relationship between the water level and the flow capacity of this device. This device can be most useful when several regulators are installed in series. The simulated effect of such a series installation is shown in Figure 13.5.

13.1.3 Operational Aspects

When the Steinscrew was developed, the goal was to have a device that would be simple to build and simple to install. It was hoped that the device would be self-cleansing and would require no more maintenance than a sewer pipe. By providing the inlet side with smooth oblique edges, most of the clogging problems appear to have been prevented. Debris that strikes the edge of the plate is moved by the flow along the edge until it passes through the space between the crown of the pipe and the regulator.

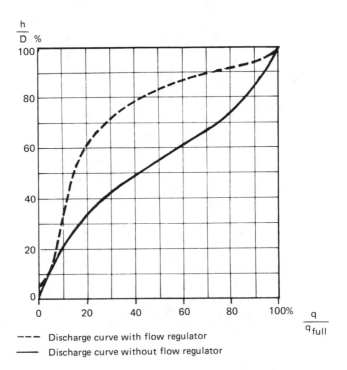

Figure 13.4 Discharge characteristics of the Steinscrew. (After Janson, Bendixen, and Harlaut, 1976.)

The regulator was tested by introducing trash and debris, some of it being pieces of wood up to $4\frac{1}{2}$ feet in length, into the water. Field (1978) reported that all debris passed through the regulators without difficulty.

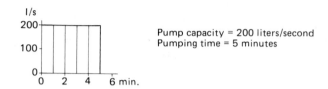

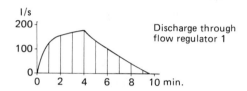

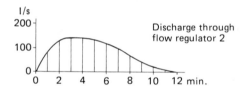

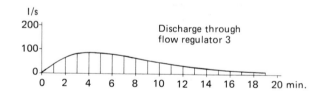

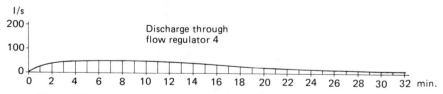

Figure 13.5 Performance of four regulators in series. (After Janson, Bendixen, and Harlaut, 1976.)

13.1.4 Hydraulics

To calculate the hydraulics of the Steinscrew, a method was developed by Schnyder and Bergerons. This method requires that the following be known:

- free water area in the upstream pipe,
- the discharge curve of the regulator in use, and
- inflow hydrograph.

The method is described by Janson, Bendixen, and Harlaut (1976). Anyone wanting to use this device should study this reference. Since the calculations are very tedious, computer modeling is suggested.

13.2 HYDROBRAKE

In the 1960s, a flow regulator called hydrobrake was developed in Denmark. This is a self-regulating device that is normally used as a flow controlling outlet at a storage facility. It does not matter if this storage is inside an in-line pipe, an underground storage vault, a surface detention basin, etc. The rate of discharge through the hydrobrake is, in part, a function of the pressure head upstream of the device.

13.2.1 Technical Configuration

The hydrobrake consists of an eccentric cylinder housing with an inlet opening located on the side. As depicted in Figure 13.6, the flow enters the hydrobrake tangentially to the outlet pipe. An outlet pipe is installed normal

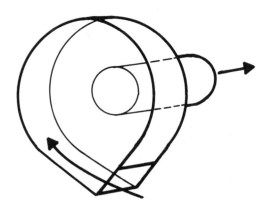

Figure 13.6 Illustration of a hydrobrake. (After Hydro-Brake Systems Inc., Portland, Maine.)

to the housing cylinder. This pipe is inserted into the basin's outlet (see Figure 13.7) using a standard O-ring to seal the annular space between the basin's outlet and the pipe. Normally, no additional anchoring is needed to hold this device in place.

The hydrobrake is manufactured from stainless steel and it can be fabricated to fit any outlet dimension. However, for very large installations it may have to be delivered in several parts and assembled at the site. It is available in two configurations illustrated in Figure 13.8. The one to the left in the figure is

Figure 13.7   Installation of a hydrobrake. (After Hydro-Brake Systems Inc., Portland, Maine.)

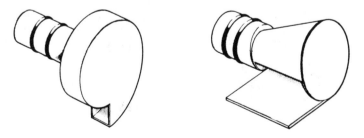

Figure 13.8   Two types of the hydrobrake. (After Hydro-Brake Systems Inc., Portland, Maine.)

a standard hydrobrake. The one on the right is designed to operate under large base flow conditions.

13.2.2 Function

As the storage basin fills, the water pressure sets the water inside the hydrobrake housing into helical motion. This causes the discharge to be significantly less than through a similar sized orifice. The available hydrostatic head upstream of the device is transformed into kinetic energy, which is only partially utilized as motion in the direction of the outlet. As can be seen in Figure 13.9, the greatest velocity component is perpendicular to the outlet.

The discharge through a hydrobrake depends on the design type and the size of the outlet. The discharge characteristics for a standard hydrobrake and an orifice, all with a 150 millimeter (approx. 6-inch) diameter outlet, are shown in Figure 13.10. Compared to an orifice, the hydrobrake has the following advantages:

1. A larger opening can be used for the same discharge rate. This can significantly reduce the risk of clogging the flow regulator.
2. The discharge varies much less with the water depth in the storage basin. As a result, a more constant release rate is obtained.

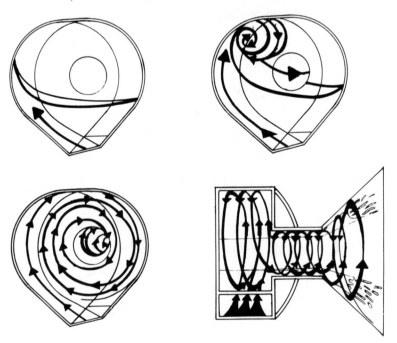

Figure 13.9 Filling of a standard hydrobrake. (After Hydro-Brake Systems Inc., Portland, Maine.)

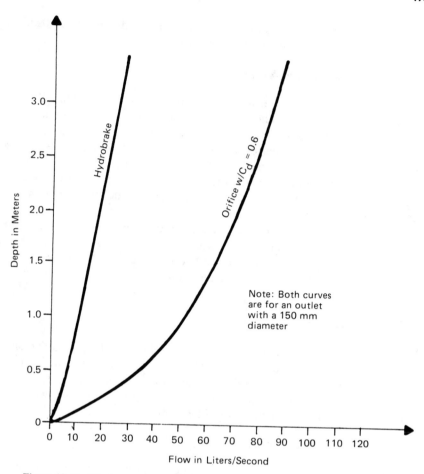

Figure 13.10 Discharge characteristics of a standard hydrobrake and an orifice with the same opening. (After Hydro-Brake Systems Inc., Portland, Maine.)

13.2.3 Hydraulics

The discharge capacity of the hydrobrake device can be calculated using the following empirical formula suggested by Hydro-Brake Systems Inc.:

$$Q = C_d \cdot A \cdot \sqrt{2g \cdot \left(h - \frac{D}{2} \right)} \tag{13.1}$$

in which Q = discharge rate,

C_d = discharge coefficient,

A = area of inlet or outlet opening, whichever is less,

D = diameter of the outlet opening,

h = water depth above invert of outlet opening, and

g = acceleration of gravity.

Discharge coefficients can be obtained from Hydro-Brake and can vary from 0.13 to 0.3 depending on the model used.

13.2.4 Operational Aspects

The hydrobrake device has been used primarily in the United States, Canada, and Scandinavia. The Danish inventor has in recent years made significant improvements to this device. The modified device is called the Mosbak regulator. The experience has been favorable, and no unusual operational problems have been encountered. There appears, however, to be some flow instability in the device when the water depth upstream of the device is relatively small. This, by itself, should not be of significant concern, since it occurs during low flows.

The only reported maintenance need was the emptying of the sedimentation pit that is usually installed just upstream of the outlet. Also, larger objects that could not pass through this device had to be removed. This must, however, be considered normal for any type of outlet.

13.3 WIRBELDROSSEL

The wirbeldrossel (i.e., turbulent throttle) was developed at the University of Stuttgart in West Germany in mid-1970s and has many similarities to the hydrobrake. As a result, the following description explains how the wirbeldrossel differs from the hydrobrake. It appears, however, that the two devices were developed totally independent of each other.

13.3.1 Technical Configuration

The wirbeldrossel has a symmetrical cylinder housing with an inlet pipe connecting tangentially to the cylinder. The outlet is a circular opening in the bottom surface of the cylinder. The opening of the outlet can be adjusted using manufactured rings of various sizes. On the opposite side from the outlet is an air supply pipe (see Figure 13.11).

This device is installed with the outlet pipe being vertical, and the device is always located in a separate chamber downstream of the storage basin. As a result, it is accessible for inspection and maintenance when the storage basin is filled (see Figure 13.12).

A modified version of the wirbeldrossel was also developed and was called wirbelvalve. This device is similar to the original wirbeldrossel, except that its rotation chamber is slanted and has a cone-shaped bottom, as illustrated in Figure 13.13. The wirbelvalve has the advantage that it requires less vertical height between the basin outlet and the downstream sewer.

The wirbeldrossel and the wirbelvalve are fabricated using welded steel plate. Corrosion protection is provided by hot-dipped galvanizing and coating

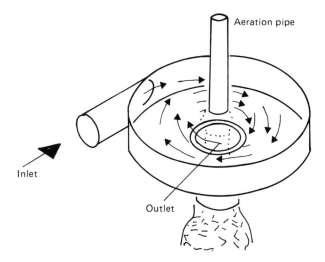

Figure 13.11   An illustration of the wirbeldrossel. (After Brombach, Umvelt und Fluidtechnik GmbH Bad Mergentheim, West Germany.)

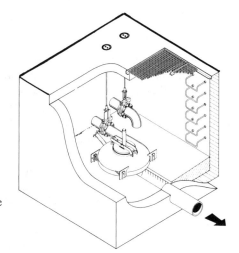

Figure 13.12   Typical installation of the Wirbeldrossel. (Umwelt und Fluidtechnik GmbH, West Germany.)

all surfaces with epoxy. Both are manufactured in standardized sizes, with inlet pipes ranging between 50- and 500-millimeters (approximately 2- to 20-inches) in diameter.

13.3.2 Function

The wirbeldrossel and the wirbelvalve are similar in their function. The flow enters the chamber tangentially, creating a rotational velocity which is proportional to the water depth upstream of the device. Due to the centripetal

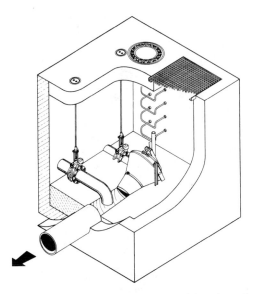

Figure 13.13   An illustration of the wirbelvalve. (Umwelt und Fluidtechnik GmbH, West Germany.)

acceleration, a pocket of air develops at the axis of rotation. This reduces the net flow area inside of the chamber, thus limiting the flow. To insure adequate air supply, an aeration pipe is connected to the axis of rotation, as shown in Figure 13.11. This aeration pipe virtually eliminates cavitation within the device, and the flow inside the cylinder has very little turbulence.

Figure 13.14 compares the capacity of a wirbeldrossel and an orifice. The capacity of the device as shown in this figure was measured at a field installation. It had a 200-millimeter (7.8-inch) inlet and a 150-millimeter (5.9-inch) outlet. Due to site conditions, the capacity of this installation was affected by backwater from downstream. Without this backwater, the flow through this device would have been reduced even further. For comparison, Figure 13.14 also contains the discharge curve for an 150-millimeter diameter orifice, which shows that the orifice has considerably larger and more variable flow capacity.

13.3.3 Operational Aspects

Because the flow inside the housing has little turbulence, even fairly large particles in the water can pass through this device without difficulty. At the Institut fur Wasserbau (1976), this device was tested for clogging by introducing various debris into the flow. The wirbeldrossel passed through sand, gravel, long rags, etc. without any problems. As far as it is known, this device has been used mainly in West Germany and Switzerland. The reported experience at existing installations has been very good.

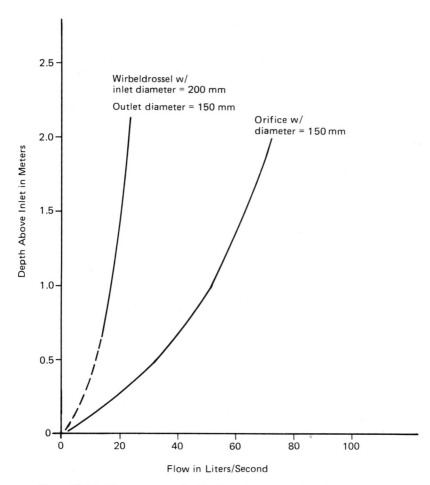

Figure 13.14 Flow capacity of wirbeldrossel vs. an orifice. (After Quadt and Brombach, 1978.)

13.4 FLOW VALVE

In the late 1970s, the, so called *flow valve* was developed in Sweden to maintain a constant release rate from a storage basin. The flow valve is a self-regulating device, and it utilizes the pressure head in the storage basin to throttle the flow. Although this device can be designed to operate in a variety of outlet conditions, we describe its installation inside a vertical inlet pipe intended to control the inflow into an overloaded sewer. By installing it at a storm sewer inlet, the inlet chamber and the street above it are used as temporary storage basins.

13.4.1 Technical Configuration

The flow valve was designed for easy installation into a circular manhole (see Figure 13.15) It is attached to the walls of the manhole, and the annular space is sealed using a rubber O-ring. The flow valve consists of a vertical central pipe surrounded by an air-filled pressure chamber, as illustrated in Figure 13.16. The top of this pressure chamber and its connection to the central pipe are made of flexible rubber fabric supported by steel housing.

13.4.2 Function

As the flow valve comes under pressure from above, the pressure is transmitted through the pressure chamber to the central pipe. The upper part of the rubber membrane is pressed in and causes the air inside the chamber to push out the rubber membrane inside the central pipe. As this occurs, the cross-section of the flow opening is reduced and the flow is throttled to maintain a constant discharge, regardless of the water depth upstream.

Figure 13.17 shows a comparison between the flow valve and an orifice. It can be seen here that as the flow increases, the flow through this device remains relatively constant. In fact, a small decrease in flow is observed with increasing depth of water upstream. The laboratory tests of this device did not, however, study it to see how it performs with trash and debris in the water.

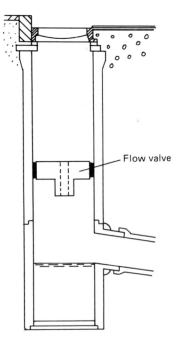

Flow valve

Figure 13.15 Flow valve installed inside a manhole. (After Bendixen and Stahre, 1983.)

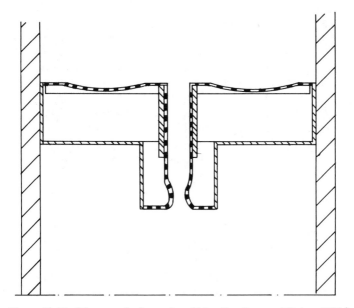

Figure 13.16   Diagram of a flow valve. (After Bendixen and Stahre, 1983.)

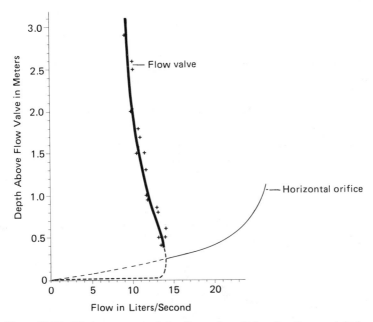

Figure 13.17   Discharge curve of the flow valve. (After Bendixen and Stahre, 1983.)

REFERENCES

Steinscrew:

FIELD, R., "Trip Report," United States Environmental Protection Agency, October, 1978.

JANSON, L. E., AND BENDIXEN, S., "New Methods for Detaining Flow Variations in Gravity Flow Lines," Stadsbyggnad No. 6, 1975. (In Swedish)

JANSON, L. E., BENDIXEN, S., AND HARLAUT, A., "Equalization of Flow Variations in a Combined Sewer," *Journal of the Environmental Engineering Division,* ASCE, December, 1976.

Hydrobrake:

BROWN, R., "Controlling Storm Water Runoff," *APWA Reporter,* May, 1978.

HYDRO-STORM SEWER CORPORATION, Miscellaneous product information from brochures, etc.

LAGER, J., SMITH W., AND TCHOBANOGOLOUS, G., "Catchbasin Technology Overview and Assessment," Environmental Protection Agency, Cincinnati Ohio, Contract 68-03-0274, 1974.

MEAGAARD, C., "A New Approach Should End Basement Floodings," *CIVIC,* April, 1979.

THEIL, P., "High Level of Flood Protection at Low Cost," APWA International Public Works Congress in Boston, October, 1978.

Wirbeldrossel:

INSTITUT FUR WASSERBAU DER UNIVERSITAT STUTTGART, "Prospect Sheet on the Wirbeldrossel," Ausgabe Bro, July, 1976. (In German)

QUADT, K. S., AND BROMBACH, H., "Operational Experiences with the Wirbeldrossel in Storm Water Overflow Basins," Korrespondenz Abwasser, No. 1, 1978. (In German)

UMWELT UND FLUIDTECHNIK GMBH, Bad Mergentheim, West Germany, Reference List on Wirbeldrossel, miscellaneous product information from brochures, catalogs, etc. (In German)

Flow Valve:

BENDIXEN, S., AND STAHRE, P., "Self-regulating Valve for Constant Flow," Vatten No. 2, 1983. (In Swedish)

14

Flow Regulation in Larger Systems

So far, we have discussed the control of flow from one basin only. However, the engineer is often faced with the possibility of a system of several storage facilities located within a separate stormwater or combined sewer network. In such cases, it is necessary to control the release rates from many storage facilities simultaneously.

The goal of controlling the storage and releases at multiple sites in a combined sewer system is to minimize the possibility of surcharged sewers while keeping the overflows to the receiving waters to a minimum. To accomplish this, the flows and water levels must be monitored simultaneously at several locations within the network. Along with this data, information about the status of various control devices in the system is transmitted to a central surveillance center. Here, the operators, often with the aid of computers, make adjustments to control settings throughout the system. This process permits a much more efficient operation of the entire system than would otherwise be possible.

14.1 BACKGROUND

Since the late 1960s, much attention has been focused on upgrading the performance of combined sewer systems in the United States and Western Europe. The upgrading usually consists of the following two actions:

- conversion of combined sewer systems into separate sewer systems; and
- improving the capabilities of combined sewer systems by the installation of storage facilities within the system.

Unfortunately, separating a combined sewer system into two separate wastewater and storm sewer systems can be extremely expensive and can rarely be justified. It is cost-effective only in a few select and densely populated areas. Also, separation of sewers was found only to marginally improve the water quality of the receiving waters over what can be done to reduce overflows through adding storage and system controls. As a result, it is possible and more cost-effective to improve the function of the existing combined sewer systems by adding detention storage to the system.

There is no standard method for upgrading combined sewer systems. Each case is unique and has to be studied for its own possibilities. Often, the resultant upgrade of the system includes a variety of actions and measures. Schilling (1986), however, after studying a large number of upgraded combined sewer systems, found certain similarities. Very often, the upgraded systems included:

1. rainfall, flow, and water level sensors;
2. pumps, inflatable weirs, and mechanical gates as flow regulators;
3. telephone lines for data and control telemetry;
4. distributed digital control systems linked to dual central minicomputers.

Table 14.1 lists several locations that utilize coordinated real time flow control as part of the system. Through the assistance of a central computer, these systems will automatically

TABLE 14.1 Several Realtime Flow Control Sewer Systems in the U.S.

Location	Description
Chicago	Detention in tunnels. Regulation with preset overflows and remote controlled pumps to feed treatment plants. Monitoring of water levels and overflows.
Cleveland	Detention in sewer network. Monitoring of precipitation, flow, and water quality.
Detroit	Detention in sewer network and storage basins. Regulation with remote controlled gates, pumps, and inflatable rubber cushions. Monitoring of precipitation and flow.
Minneapolis-St. Paul	Detention in sewer network. Regulation with remote controlled gates and inflatable rubber cushions. Monitoring of precipitation, flow, and water quality.
San Francisco	Detention in sewer network, tunnels, and storage basins. Monitoring of precipitation and flow.
Seattle	Detention in sewer network. Regulation with remote controlled gates and pumps. Monitoring of precipitation, flow, and water quality.

- open or close gates and valves,
- raise or lower overflow weir crests, and
- turn pumps on or off.

Figures 14.1 and 14.2 illustrate two examples of coordinated flow regulating structures used in larger systems. The first one, located in Seattle, Washington uses available volume in an existing combined sewer. Here the flow is normally routed to a trunk sewer which then conveys it to the treatment plant. As the flow increases, and the control system determines a need, the control gate throttles the flow to the trunk sewer and causes the excess water to be stored. Should the flow exceed storage capacity, it then overflows into the receiving waters. The second system, located in Minneapolis-St. Paul, although utilizing different equipment has almost an identical operational scenario.

14.2 DESIGN CONSIDERATIONS FOR A REGULATING SYSTEM

Usually the primary goal is to improve water quality by optimizing the use of existing sewer networks. The objectives include the reduction of overflows to the receiving waters and the equalization of flow to the treatment plant. These tasks require considerable sophistication in that:

- All flow regulators in the sewer network need to receive continuous control signals during the entire storage and detention cycle.

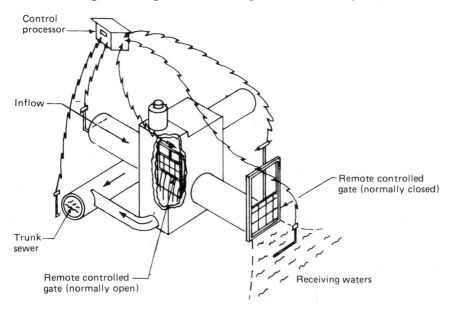

Figure 14.1   Flow regulating structure in Seattle. (After Leiser, 1974.)

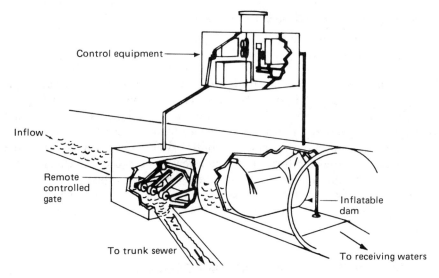

Figure 14.2 Regulating structure in Minneapolis-St. Paul. (After Anderson and Callery, 1974.)

- The signals from each site have to transmit information about storage and flow at each of the many flow regulators within a catchment.
- The control of each regulator must also consider how all of the regulators work together as a system.
- The processing of all data received at the control center and its transformation into control signals has to occur almost instantaneously.
- The runoff hydrograph needs to be anticipated to optimize the storage resources. This has to be done with simple models, as there is not enough time for complete simulation of all dynamic conditions.
- The computer hardware and software need to accommodate growth and changes in equipment, sewer network, treatment plant capacity, population growth, and treatment technology.

In Figure 14.3, the operation of a single catchment is illustrated. The inflow into the storage basin depends on the character of the drainage area, the sewer network, and the temporal and spatial distribution of the precipitation. Flow regulation is determined by how the inflow is to be distributed between storage, spillage, and transport downstream (see Figure 14.4). This cannot be decided without understanding downstream system capacities, existing and projected flow conditions, and how the flows from other subbasins will interact.

Large sewer networks consist of smaller networks serving individual catchments (see Figure 14.5). Here, several smaller networks are connected to

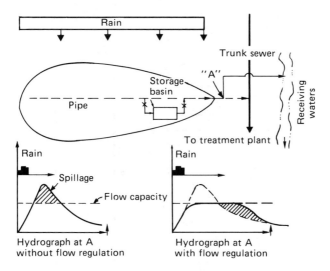

Figure 14.3   Regulation of flow in a catchment within a larger sewer network. (After Grigg and Labadie, 1974.)

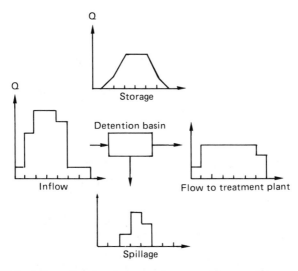

Figure 14.4   Inflow allocation between storage, spillage, and transport downstream. (After Bruce and Bradford, 1977.)

a trunk sewer, which conveys flow to a treatment plant. By knowing the conditions within each catchment, it is possible to coordinate various regulators in a manner that optimizes the performance of the entire system. Of course, optimization will be determined by the goals and objectives of the system. In this example, the goals could include

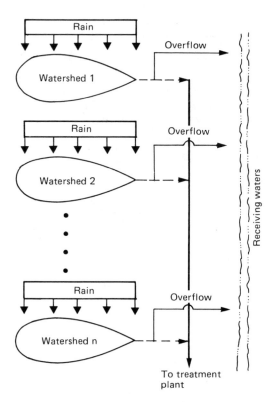

Figure 14.5 Composition of a larger sewer network. (After Grigg and Labadie, 1974.)

- The flow to the treatment plant is to be kept uniform.
- Wastewater from certain areas is to receive priority transport to the treatment plant.
- Spillage of untreated wastewater is to be minimized.
- Spillage is to occur only at certain locations.

Coordinated flow regulation cannot be achieved using only information about flows and volumes. Effective flow regulation also requires the collection of precipitation data throughout the watershed and the prediction of how much runoff may occur, where it will occur, and how it will be distributed over time. In some cases, it may even be necessary to predict rainfall in advance of actual measurement. Obviously, such systems are expensive and difficult to design, install, and operate, and as a result their use to date has been limited to larger cities.

14.3 EXAMPLE: BREMEN, GERMANY CONTROL SYSTEM

A computer control system for the sewer network of the city of Bremen, West Germany was installed in the late 1970s. One such installation was in a watershed having an area of 1,550 acres (628 ha) located on the west bank of the river Weser. The older city within this watershed had combined sewers, while the newer fringe zones had separated sewer systems. The stated goal for this project was to control storm runoff such that:

- Backing up of stormwater within the sewer network would be eliminated.
- Spillage of untreated combined sewage into receiving waters would be significantly reduced.
- Operational reliability would be improved.
- Energy costs would be reduced.
- The operational environment for personnel would be improved through automation.
- The use of existing treatment facilities would be optimized.

Martin et. al. (1978) described this system in detail. Following is a brief description of some of the components of the Bremen system.

14.3.1 The West Bank Pumping Station

The West Bank pumping station shown in Figure 14.6 is the main control center of the Bremen system. Data from three rain gauges, 19 pumping stations, and 17 water level meters are received and processed at this location. The central computer then sends control signals to all the pumping stations and storage facilities within the watershed.

The West Bank pumping station is equipped with trash racks and grit chambers that remove the coarse particles from the flow before it is pumped to the treatment plant. The station's maximum pumping capacity is rated at 88 cubic feet per second (2.5 m³/s), which is four times the dry weather flow. When inflow exceeds this capacity, the excess water is transferred to storage basins having a total capacity of 353,000 cubic feet (10,000 m³). This storage capacity was doubled since original construction to improve treatment efficiencies at the downstream wastewater treatment plants.

Combined wastewater is routed to the storage basins by three screw pumps having a total capacity of 320 cubic feet per second (9 m³/s). The storage basins, in turn, are emptied into the pumping station by gravity after the storm flows subside. When the storage basins are full and the incoming flow still exceeds the pumping capacity of the station, the excess water overflows into the Weser River.

Figure 14.6 West Bank pumping station and control building in Bremen, West Germany.

14.3.2 The Krimpel Pumping Station

This pumping station is the second largest pumping station in the system. It has a pumping capacity of 25 cubic feet per second (0.7 m^3/s). Like the West Bank station, it also has a storage capacity of 353,000 cubic feet (see Figure 14.7). This station was designed to handle a one-year rainstorm, and the station is expected to have, on the average, a spill once a year.

14.3.3 Measuring Equipment

The equipment for data collection within this system is varied and complex, and we touch only briefly on some of the equipment in use. For example, the rain gauges, which telemeter data to the West Bank pumping station, were especially developed for this project. The commercial gauges available when this system was designed (i.e., early 1970s) were judged by the designers to be unreliable and incompatible with the central processing system. Since then, many new products have been introduced into the market that probably could do the job without modification.

As mentioned earlier, the water level is measured at 17 locations within the watershed. Six of these locations measure a differential in water levels, thereby providing data for the estimate of flow rates. The other gauges report water levels at overflows and storage basins. In almost all cases, pressure transducers were installed in conventional manholes, as illustrated in Figure 14.8.

Figure 14.7   Storage basin at Krimpel pumping station in Bremen, West Germany.

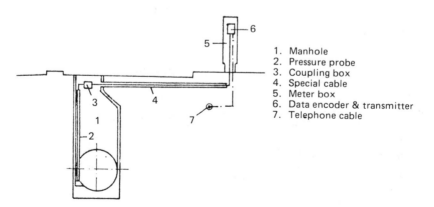

1. Manhole
2. Pressure probe
3. Coupling box
4. Special cable
5. Meter box
6. Data encoder & transmitter
7. Telephone cable

Figure 14.8   Water level measurement installation in Bremen. (After Martin et. al., 1978.)

REFERENCES

ANDERSON, J., AND CALLERY, R., "Remote Control of Combined Sewer Overflows," *Journal of WPCF*, No. 11, November, 1974.

BRUCE AND BRADFORD, "Optimal Storage Control in a Combined Sewer System," *Journal of the Water Resources Planning and Management Division*, WR1, May, 1977.

FAHRNER, H., "Planning of Storm Basin Discharge Control Using the 'Automatic' Throttle plate," Abwassertechnik No. 5, 1979. (In German)

FLOOD CONTROL COORDINATING COMMITTEE, *Development of a Flood and Pollution Control for Chicagoland Area*, Summary Reports, August, 1972.

GRIGG, N., LABADIE, J., SMITH G., HILL, D., AND BRADFORD, B., "Metropolitan Water Intelligence Systems," *Completion Report, Phase II*, Department of Civil Engineering, Colorado State University, June, 1973.

GRIGG, N., AND LABADIE, J., "Computing the Big Picture," *Water and Wastewater Engineering*, May, 1975.

GRIGG, N., LABADIE, J., AND WENZEL, H., "Metropolitan Water Intelligence Systems," *Completion Report Phase III*, Department of Civil Engineering, Colorado State University, July, 1974.

LABADIE, J., GRIGG, N., AND BRADFORD, B., "Automatic Control of Large-Scale Combined Sewer Systems," *Journal of Environmental Engineering*, EEI, February, 1975.

LAGER, J., "Control of Dry Weather Pollution," *Water and Wastewater Engineering*, September, 1974.

LARGER, J., SMITH, W., LYNARD, W., FINN, R., AND FINNERMORE, J., "Urban Stormwater Management and Technology," *Update and Users's Guide*, EPA-600/8-77-014, July, 1977.

LEISER, C.P., "Computer Management of a Combined Sewer System," EPA-679.274-022, July, 1974.

MARTIN, G., MECKELBURG, H. J., VOUGT, D., AND WINTER, J., "Central Data Recording Systems for Controlling the Municipal Sewer Net of the City of Bremen," Korrespondenz Abwasser No. 3 1978. (In German)

MUNICIPALITY OF METROPOLITAN SEATTLE, "Maximizing Storage in Combined Sewer Systems," EPA 11022 ELK 12/71.

PARTHUM, "Building for the Future, the Boston Deep-Tunnel Plan," *Journal WPCF*, April, 1970.

SCHILLING, W., "Urban Runoff Quality Management by Real-Time Control," *Urban Runoff Pollution*, Springer-Verlag, Berlin, 1986. (In English)

TROTTA, P., LABADIE J., AND GRIGG, N., "Automatic Control Strategies for Urban Stormwater," *Journal of Hydraulics Division*, HY 12, December, 1977.

WESTON, R., "Combined Sewer Overflow Abatement Alternatives," EPA 11024 EXF 08/70, Washington, D.C., 1971.

15

Basic Principles

15.1 GENERAL

The term *storage facility* is used here to describe various types of facilities for the retardation of separate stormwater or combined stormwater–wastewater. The technical configuration of these facilities is subordinate to their intended function, of which there are three basic types:

- infiltration and percolation facilities,
- detention facilities, and
- retention facilities.

15.2 INFILTRATION AND PERCOLATION FACILITIES

As described in Part 1, the storage volume for an infiltration facility is over the infiltrating surface. The storage volume in a percolation basin consists of the pore volume of the stone filling. These storage facilities are emptied either through percolation to the underlying layers of soil or through specially designed underdrain pipes. They can also be equipped with an overflow outlet to drain off the excess water (see Figure 15.1).

The geohydrologic conditions at the site have to be known and understood in order to design infiltration and percolation facilities. Data on soil permeability and porosity, groundwater level and its fluctuations with seasons, soil profiles, etc. are needed for proper design.

15.3 DETENTION FACILITIES

Detention facilities store water for a relatively short period of time and, unless they are used for stormwater quality enhancement, they are primarily used to reduce the peak rate of flow downstream. When the detention facility is connected in-line with the conveyance system, as shown in Figure 15.2, all flows entering the basin are metered through the outlet and are temporarily stored within the basin. When the volume of water exceeds the available storage volume, the excess water is released via an emergency spillway.

If a detention basin is connected off-line to the conveyance system, all initial stormwater flows and dry weather flows bypass the basin. Flow enters the basin only after a preset rate is exceeded, at which point the flows exceeding the preset rate are diverted into the pond. As a result, only the "peak" portion of the hydrograph is stored in the pond.

Obviously, detention facilities cannot be sized to handle all possible storm scenarios, and often a single or a set of several design storm frequencies is selected for sizing purposes. Chapters 18 and 19 discuss the design principles for the sizing of detention facilities.

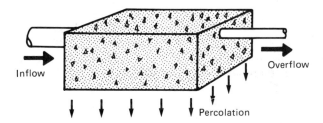

Figure 15.1   Schematic of a percolation facility with an overflow.

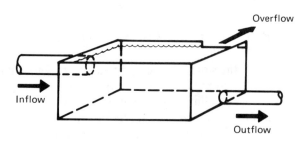

Figure 15.2   An in-line detention facility with a spillway.

15.4 RETENTION FACILITIES

Retention facilities trap the water for an indefinite period of time. As a result, they are often used as sedimentation basins to remove some of the pollutants from the stormwater. The degree of pollutant removal is a function of

- quality of inflow, and
- holding time in the storage basin.

Unlike infiltration/percolation or detention facilities, retention facilities do not typically have an operational bottom outlet. It is a good idea, however, to provide a bottom drain, as shown in Figure 15.3, for maintenance purposes. In operation, a retention basin is constantly full of water. A detailed discussion of how to size a retention facility for water quality enhancement appears in Part 4 of this book.

15.5 REGIONAL APPROACH TO ON-SITE DETENTION

Although much work has been done on the sizing and the effectiveness of individual detention basins using various design frequency scenarios, very little has been reported on how a random system of individually sized detention basins affect the flow regime along the drainageways. McCuen (1974) reported a computer model study of the Gray Haven Watershed using basin information reported by Tucker (1969). The study modeled 17 subbasins and two scenarios of detention storage. The first scenario used 12 ponds (total storage was not reported by McCuen), and the second used 17 ponds with 22,000 cubic feet of total storage. The watershed had an area of 23.3 acres and had 52% impervious surface. McCuen suggested on the basis of the model study

(1) That the "individual-site" approach to stormwater detention may actually create flooding problems rather than reduce the hydrologic impact of urbanization; and (2) that a regional approach to urban stormwater management may be more effective than the "individual-site" approach.

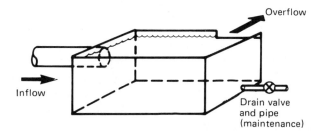

Figure 15.3   A retention facility with a spillway.

Hart and Burges (1976) investigated a hypothetical 2,000 acre watershed by modeling three subbasins with three individually sized detention facilities. They concluded the following:

> Restricting the outflow from a retention facility to a level less than the undeveloped rate could achieve a composite peak flow rate that would equal the preurbanized flow but would run at a much greater duration at that rate. The increased flow duration would have potentially undesirable effects on the channel system.

In 1981–1982 Urbonas and Glidden (1983) studied the potential effectiveness of various detention policies. They modeled a 7.85 square-mile watershed in the Denver area under preurbanization and posturbanization conditions. The kinematic model had 56 subbasins and 52 channel segments. After determining the "historic" and fully developed runoff at various points in the watershed, the model was modified to incorporate 28 individually sized detention basins. Computer runs were made using the 2-, 10- and 100-year design storms and three recorded storms.

Next, using the results of these runs, Urbonas and Glidden tested several empirical on-site detention sizing equations for the study watershed. These equations were limited to the sizing of detention basins to limit peak runoff from the upstream watershed to the historic 10- and 100-year runoff peaks. The storage–discharge characteristics of the previously modeled 28 detention basins were modified to conform with these empirical equations and the system response was again tested. They concluded the following:

1. The peak flows along the drainageways can be controlled by random on-site detention to almost the "historic" levels only for the large intensity storms such as the 10- and the 100-year design storm.

2. The control of runoff peaks from these large storms is limited to the design frequency used to design the detention basins. The peak flows from other frequency storms are either not controlled, or controlled only marginally. In other words, a system of on-site ponds designed to control the 10-year flood, if fully and properly implemented, has the potential of controlling the peaks along the urban drainageways resulting from the 10-year storm. It will not do the same for lesser or larger storms.

   Detention basins designed to control both the 10- and 100-year peaks performed better in controlling a range of larger runoff events than basins designed to control only a single runoff frequency.

3. A system of detention basins designed to control more frequent storms, such as the 2-year or lesser storm, are effective in controlling peaks from the frequently occuring storms only immediately downstream of each storage facility. Control of frequently occuring runoff peaks along the downstream drainageways becomes ineffective as the number of detention basins increases.

4. It appears feasible to develop simple empirical on-site detention sizing equations for a given watershed and possibly for an entire metropolitan area. The simulations by Urbonas and Glidden showed that a system of on-site detention basins designed using the simplified equations can be as effective as a system of basins designed individually.

The authors believe that a system of detention basins sized using uniform volume and release requirements can, in fact, be more effective in controlling peaks along the drainageways than a system of random designs. The latter suffers from the fact that each detention basin design can vary in size and discharge characteristics due to the approach, assumptions and expertise of each designer.

REFERENCES

McCuen, R. H.,"A Regional Approach to Urban Stormwater Detention," *Geophysical Research Letters,* 74–128, pp. 321–22, November, 1974.

Tucker, L. S.,"Availability of Rainfall-Runoff Data for Sewered Drainage Catchments," ASCE Urban Water Resources Research Program Technical Memorandum No. 8., New York, 1969.

Urbonas, B. R., Glidden, M. W.,"Potential Effects of Detention Policies," Proceedings of the Second Southwest Regional Symposium on Urban Stormwater Management, Texas A & M University, November 1983.

16

Precipitation Data Needs for Estimating Storage Volumes

16.1 GENERAL

When an engineer designs a storm sewer or an open channel, the main concern is the rate of rainfall runoff (i.e., peak flow) the conveyance facility needs to handle. Flows from storms with less rainfall intensity than designed for are safely conveyed within the facility without backing up water into streets, parking lots, basements, etc. The volume of runoff is only of secondary interest.

Storage facilities are used in storm drainage systems to either extend system capacities, to provide flow equalization, or to provide water quality enhancement. Design of storage, however, requires knowledge of rainfall and runoff volumes. Rainfall intensity, by itself, is not sufficient information. As a result, rainfall information that is adequate for the sizing of storm sewers is not adequate for the sizing of storage facilities.

We now discuss various types of precipitation data that may be considered or acquired for the design of storage facilities. We have classified the data into the following groups:

- intensity–duration–frequency data;
- standardized design storms;

- chronologic rainfall records;
- snow melt; and
- chronologic series of flow records.

The first two groups currently form the basis for most drainage and water quality design in the United States and Europe. Although often criticized by some, they are well entrenched into engineering practice, offer fast results, and are relatively simple to use.

As the designs become more complex and system-operations-oriented, it becomes necessary to shift to the use of chronologic rainfall records. Unfortunately, these may not be available to many communities. Data from other, hopefully meteorologically similar, sites has to be borrowed whenever local records are inadequate, thereby reducing confidence in the data's applicability at the design site.

Also, the analysis or operation of large systems require not only the knowledge of how precipitation varies with time, but also how it is distributed over the watershed at any given time. Precipitation data of this complexity is practically nonexistent and, therefore, storm runoff storage and conveyance system designs are often based on less than complete information. This fact should not deter the engineer from doing his or her job, however. Generalized information, although not as complete as we often would like to have, can serve as an adequate basis for design. What's important is that we recognize the limitations of the information and use it to the best of our abilities.

16.2 INTENSITY–DURATION–FREQUENCY DATA

The use of rainfall intensity–duration–frequency data became a part of the storm sewer design practice when the Rational Formula was introduced. Although the Rational Formula is used primarily for the sizing of storm sewers, it is currently also being used for the design of detention, retention, infiltration, and percolation facilities. The Rational Formula assumes that the effective rainfall intensity over the entire catch-basin is equal to the intensity found at the time of concentration. Figure 16.1 illustrates this principle for a five-year storm.

Most of the rainfall data for the development of the intensity–duration–frequency (i.e., I-D-F) curves in the United States are collected by the U.S. Weather Service, primarily at its first-order stations. However, the users of these data need to understand that each data set is representative only of the rainfall characteristics at the gauging site.

The I-D-F curves are constructed by searching the records of all storms for the most intense periods of rainfall. Rainfall depths are then ranked and tabulated by duration, as shown in Table 16.1. The tabular values are then converted to I–D–F graphs. Another example of a set of I-D-F values, for a

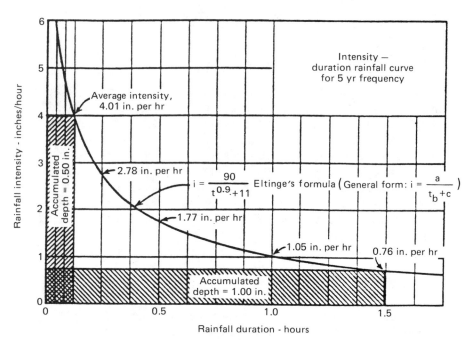

Figure 16.1   Intensity–duration curve for a five-year storm.

TABLE 16.1   Intensity–Duration–Frequency; Chicago, Ill.

Rank	RETURN PERIOD IN YEARS		MAXIMUM DEPTH IN INCHES AND STORM NUMBER FOR STATED DURATION			
	Annual	Exceedance	5-min.	10-min.	15-min.	25-min.
1	36.0	35.0	.61(7)	1.11(7)	1.29(7)	1.64(7)
2	18.0	17.5	.58(8)	.96(5)	1.22(5)	1.58(4)
3	12.0	11.7	.55(6)	.94(6)	1.16(6)	1.49(3)
4	9.0	8.8	.53(1)	.92(8)	1.16(4)	1.45(2)
5	7.2	7.0	.51(5)	.88(4)	1.15(8)	1.45(8)
6	6.0	5.8	.50(4)	.80(3)	1.12(3)	1.39(5)

Storm Number	Storm's Date
(1)	June 29, 1920
(2)	July 7, 1921
(3)	August 11, 1923
(4)	June 20, 1928
(5)	August 11, 1931
(6)	June 26, 1932
(7)	September 13, 1936
(8)	July 6, 1943

storm having a much longer duration, are listed in Table 16.2. As can be seen by these two examples, rainfall I–D–F information can be radically different, depending on the local hydrologic conditions and the purpose for which this information will be used.

Examination of Table 16.1 reveals that the six rankings of I–D–F values for four durations of rainfall came from eight different storms. For example, the fourth ranked event had its 5-minute rain depth from storm (1), its 10-minute depth from storm (8), its 15-minute depth from storm (4), and its 20-minute depth from storm (2). This approach of normalizing rainfall data maximizes the depth at each duration. It does not consider if the successive rainfall depths are from the same storm, or even from the same type of storms. Nevertheless, despite this apparent inconsistency in the construction of I–D–F curves, this technique permits engineers to come up with reasonable rainfall values for estimating storage volumes.

16.3 STANDARDIZED DESIGN STORMS

Another popular approach for sizing storage facilities is to use standardized design storms. In practice, design storms are developed in a variety of ways. Some are derived using the I–D–F information, while others are derived using other available rainfall data. A design storm, in theory, is expected to be representative of many recorded rainstorms and may also be expected to reflect the intensity, volume, and duration of a storm having a given recurrence frequency (i.e., 5-year, 50-year, etc.).

When a standardized design storm is used to design a storage facility, it is assumed that the facility will operate at capacity, on the average, at the same recurrence frequency as the design storm. This assumption presupposes many things.

First, it assumes that the design storm has a rainfall volume equivalent to a real storm having the same statistical recurrence frequency. Second, it assumes that the temporal distribution of rainfall within the storm is represen-

TABLE 16.2   Intensity–Duration–Frequency, Stockholm, Sweden

RETURN PERIOD IN YEARS	MAXIMUM DEPTH IN INCHES FOR STORMS OF FOLLOWING DURATIONS (DAYS)			
	0.5	1.0	2.0	4.0
0.25	0.51	0.62	0.76	0.92
0.50	0.68	0.83	1.01	1.22
1.00	0.87	1.06	1.28	1.56
2.00	1.09	1.32	1.60	1.94
5.00	1.46	1.78	2.16	2.62
10.00	1.86	2.28	2.75	3.33

tative of a storm occurring in nature. Third, it assumes that the storage facility is empty and ready to accept all of the runoff when the design storm occurs. Namely, the storage basin is not partially full from runoff that may have occurred only a few hours or a few days prior to this storm. Fourth, the design storm is uniformly distributed over the entire watershed. This is probably a reasonable assumption for small watersheds, but is not appropriate for the larger watersheds.

Standardized design storms are used extensively in practice. In fact, for practical reasons their use may be the only readily available alternative at many locations. Nevertheless, the use of standardized design storms for the design of storage facilities has been criticized by many investigators (Marsalek, 1978; Marsalek et.al., 1986; McPherson, 1976, 1977; Walesh, 1979; Wenzel, 1978). Most of the criticism centers around the argument that the use of design storms cannot adequately reproduce the runoff volume frequency distribution (see Figure 16.2). Although this criticism appears to have merit, often the lack of local rainfall data still mandates that the storage facilities be designed using standardized design storms. We now describe some of the design storms being used in the United States and in Europe.

16.3.1 Block Rainstorm

The simplest form of standardized design storm is the so-called *block rainstorm*. As is seen in Figure 16.3, a block rainstorm has a constant intensity during the entire event. The intensity of the block rainstorm, for different

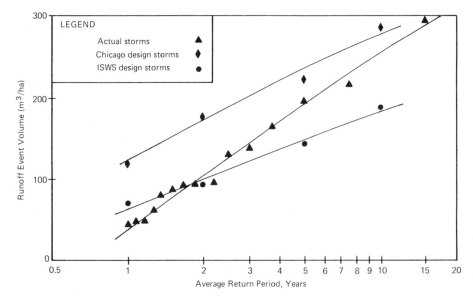

Figure 16.2 Runoff volumes resulting from actual and two design storms. (After Marsalek, 1978.)

durations and recurrence frequencies of rain, can be obtained directly from the I–D–F curves, as illustrated in Figure 16.4. It is clear that the development of the block rainstorm has its roots in the Rational Method.

As illustrated in Figure 16.3, the block rainstorm represents only the average of the most intense portion of an actual storm. The rain that falls before or after this period is not included. The use of the block rainstorm may have merit in the sizing of storm sewers, but its use is of questionable validity in the sizing of storage facilities. It is expected that the use of the block rainstorm will result in too small of a storage volume.

16.3.2 Sifalda and Arnell Rainstorms

A modification to the block rainstorm was suggested by Sifalda (1973). The modification includes a trapezoidal rainfall pattern before and after the block rainstorm, as illustrated in Figure 16.5.

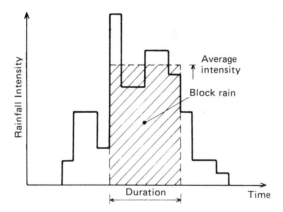

Figure 16.3   Definition of the block rainstorm as a design storm.

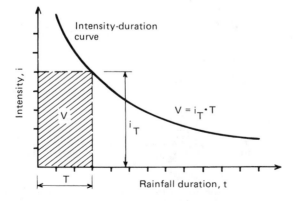

Figure 16.4   Derivation of the block rainstorm using the I–D–F curve.

Arnell (1978), after comparing rainfall duration–volume statistics, concluded that the Sifalda rainfall tended to overstate the rainfall volume. (His findings are shown in Figure 16.6.) By modifying the Sifalda rainstorm, Arnell succeeded in constructing a design rainstorm that, at least for Sweden, appears to be consistent with duration–volume statistics. Arnell's storm is illustrated in Figure 16.7. This approach appears to have achieved volumetric

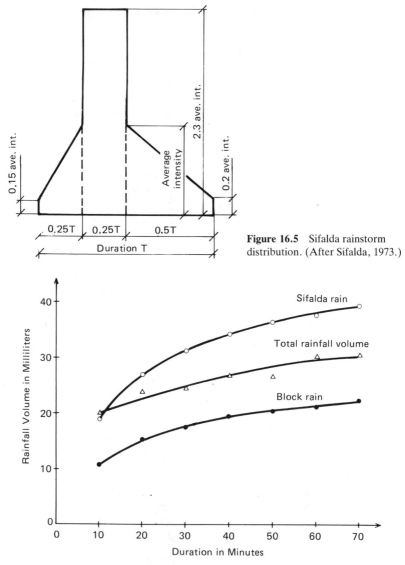

Figure 16.5 Sifalda rainstorm distribution. (After Sifalda, 1973.)

Figure 16.6 Duration–volumes of Block and Sifalda rainstorms. (After Arnell, 1978.)

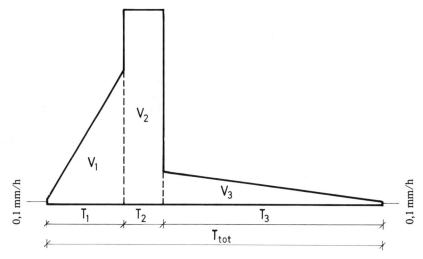

Figure 16.7 Arnell rainstorm distribution. (After Arnell, 1982.)

consistency; however, it has not been tested against the probabilities of storage basin performance using a continuous rainfall record.

16.3.3 Chicago and the ISWS Design Rainstorms

Keifer and Hsien Chu (1957) described the development of the now well-established Chicago design rainstorm. This method has been widely used in North America because it is simple to derive using standard I–D–F information. Its simplicity is also its major shortcoming; namely, it is merely a redistribution of the I–D–F.

The developers of the Chicago rainstorm attempted to incorporate some of the features found in actual rainstorms. The storm takes into account the maximum rainfall intensities of individual storms, the average antecedent rainfall before the peak intensity, and the relative timing of the peak. To develop the Chicago design rainstorm, one needs to study a number of recorded storms to determine the value of t_r such that

$$t_r = \frac{t_p}{T} \tag{16.1}$$

in which t_r = ratio of time from start of storm to peak intensity to total duration of the storm (average of recorded storms),

t_p = time from start of storm to peak intensity, and

T = total duration of the storm.

The hyetograph is then constructed using the following formula:

$$i_{av} = \frac{a}{t_d^b + c} \tag{16.2}$$

in which i_{av} = average rainfall intensity over duration t_d,
 t_d = duration of storm for the average intensity, and
a, b, c = constants used to fit the data.

After studying numerous recorded rainstorms in Illinois, Huff (1967) suggested an alternate method for developing design storms in Illinois. This technique was later incorporated by Terstriep and Stall (1974) into the Illinois Urban Area Simulator (i.e., ILLUDAS), after which it became known as the Illinois State Water Survey (i.e., ISWS) design storm. In this procedure, the one-hour rainfall depths are distributed into a hyetograph of desired duration using a normalized relationship developed from Illinois data. Marsalek (1978) compared these two procedures, and his findings for the prediction of runoff peaks are presented in Figure 16.8. The actual rainfall patterns are compared in Figure 16.9.

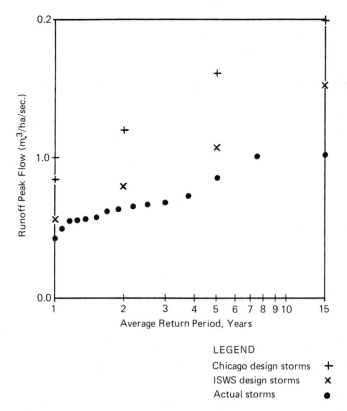

Figure 16.8 Comparison of runoff peaks for Chicago and ISWS design storms vs. recorded storms. (After Marsalek, 1978.)

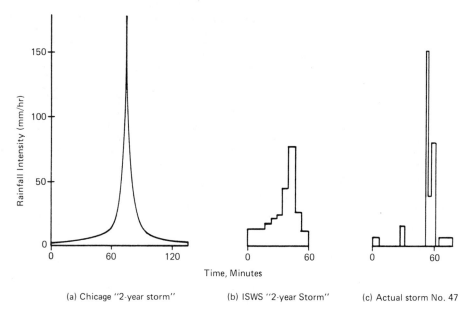

(a) Chicage "2-year storm"   (b) ISWS "2-year Storm"   (c) Actual storm No. 47

Figure 16.9   Comparing Chicago, ISWS, and actual hyetographs. (After Marsalek, 1978.)

16.3.4 Colorado Urban Design Storms.

As part of an effort to update the Colorado Urban Hydrograph Procedure for the Denver area, Urbonas (1979) developed a regional set of normalized design storm distributions. These design storms were based on a runoff and rainfall data base within the semi-arid region of Denver. For watersheds of 5 square miles or less, a Colorado Urban Design Storm hyetograph can be obtained for recurrence frequencies of 2, 5, 10, 50, or 100 years by distributing the one-hour depth. This is done using a given multiplier for each of the five-minute increments within the two-hour storm. The distribution multipliers for each reoccurrence storm are given in Table 16.3, which was taken from the *Urban Storm Drainage Criteria Manual* (1969). Good comparisons of the calculated peak runoff rates were found when tested against simulation results obtained using long-term series of recorded rainfall records (see Figure 16.10).

16.4 CHRONOLOGIC RAINFALL RECORDS

Since the occurrence of precipitation is a stochastic process, the occurrence of the resultant runoff is also a stochastic process. Besides precipitation, runoff is affected by the watershed characteristics and the configuration of the convey-

TABLE 16.3. Design Storm Distribution of 1-hour NOAA Atlas Depth
for Denver, Colorado Area

Time Minutes	PERCENT OF 1-HOUR NOAA RAINFALL ATLAS DEPTH				
	2-year	5-year	10-year	50-year	100-year
5	2.0	2.0	2.0	1.3	1.0
10	4.0	3.7	3.7	3.5	3.0
15	8.4	8.7	8.2	5.0	4.6
20	16.0	15.3	15.0	8.0	8.0
25	25.0	25.0	25.0	15.0	14.0
30	14.0	13.0	12.0	25.0	25.0
35	6.3	5.8	5.6	12.0	14.0
40	5.0	4.4	4.3	8.0	8.0
45	3.0	3.6	3.8	5.0	6.2
50	3.0	3.6	3.2	5.0	5.0
55	3.0	3.0	3.2	3.2	4.0
60	3.0	3.0	3.2	3.2	4.0
65	3.0	3.0	3.2	3.2	4.0
70	2.0	3.0	3.2	2.4	2.0
75	2.0	2.5	3.2	2.4	2.0
80	2.0	2.2	2.5	1.8	1.2
85	2.0	2.2	1.9	1.8	1.2
90	2.0	2.2	1.9	1.4	1.2
95	2.0	2.2	1.9	1.4	1.2
100	2.0	1.5	1.9	1.4	1.2
105	2.0	1.5	1.9	1.4	1.2
110	2.0	1.5	1.9	1.4	1.2
115	1.0	1.5	1.7	1.4	1.2
120	1.0	1.3	1.3	1.4	1.2
Totlas	115.7	115.7	115.7	115.6	115.6

After *Urban Storm Drainage Criteria Manual*

ance and storage system. As a result, the statistical distribution for rainstorms may not be the same as for the resultant runoff.

If a chronologic record of rainfall is used to size storage facilities, it first needs to be transformed into a record of runoff. This can be done with the aid of computer runoff simulation models, and the results can then be statistically analyzed. The assumption that the runoff and the rainfall, as is the case with design storms, have the same statistical properties no longer is an issue when using the chronological record. This is an advantage, since the conveyance system and storage within this system can strongly influence the runoff process.

Unfortunately, the cost of simulating a continuous long period of runoff under different system configurations is time-consuming and, as a result, expensive. This type of analysis can be justified only for large systems, where the numbers of combined wastewater–stormwater overflows are limited by an NPEDS permit, and where real time operations are to be used in combined sewer systems.

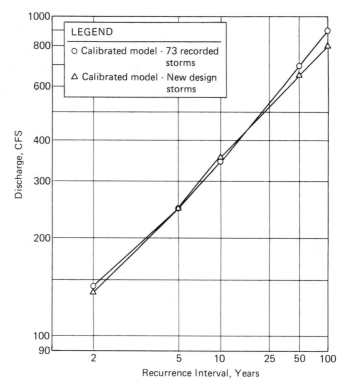

Figure 16.10   Runoff estimates using Colorado design storms vs. recorded storms. (After Urbonas, 1979.)

Walesh (July, 1979) proposed an alternative to continuous long-term simulation. He suggested that a chronologic rainfall record be analyzed to develop an annual series of peak flows and runoff volumes for a select variety of small watersheds. With the aid of this information, several representative storms that array around the desired T-year recurrence period are selected. The selected storms, along with their antecedent rainfall, can then be used in the design or evaluation of storm sewers, channels, and detention storage facilities. Thus, instead of simulating the runoff for, say, all storms in a 20-, 40- or 60-year period, one needs to only simulate maybe 30 to 50 storms in order to develop T-year recurrence period distribution of runoff peaks or volumes.

16.5 SNOW MELT

Stormwater detention storage facilities are typically designed to handle the runoff from rainstorms. However, with the advent of water quality concerns for urban runoff, it may, in some cases, be necessary also to account for snow melt. Unlike the intensity–duration–frequency data for rainfall, there is no equivalent methodology for estimating runoff due to snow melt. There are, however, procedures available for estimating the rates of snow melt that are based on local experiences.

In order to calculate the rate of runoff from snow melt, one must first be able to determine the intensity of snow melt. Snow melting in spring usually occurs slowly at first, accelerating as the snow becomes water-soaked and the albedo of the snow is reduced. In Figure 16.11 we see the intensity increasing as the seasons shift from winter to spring in the month of April. As the warm weather advances, the intensity of the snow melt and runoff increases. At the same time this is happening, the amount of snow in the fields decreases. Eventually, the mass of the snow shrinks such that, despite increasing snow melt rates, the resultant runoff begins also to decrease.

Theoretically, the snow melt intensity could reach about 0.29 inches per hour (7 mm/hr). In practice, however, the maximum intensity rarely exceeds one-half of this rate. When calculated for a whole day, heavy snow melt can produce 0.82 to 1.1 inches of water.

Because the ground is usually frozen, one would expect most of the snow melt to run off. However, it has been found that much of the snow melt actually infiltrates through the frozen ground. As in unfrozen ground, the infiltration rate decreases as the soil becomes saturated. What often happens

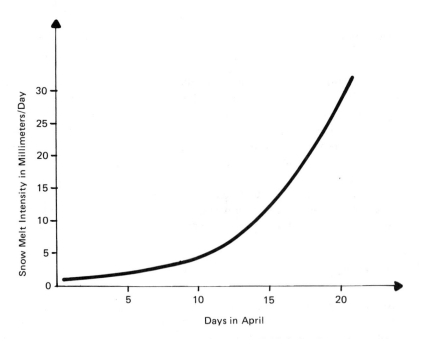

Figure 16.11   Snow melt rate in Sweden in an open field. (After Bengtsson et. al., 1980.)

is that when the snow melt rate is the highest, the ground is saturated and most of the water becomes surface runoff.

Snow melt generally does not control the size of conveyance facilities for stormwater systems. Nor does snow melt control the sizing of conveyance or treatment facilities for combined systems. It is usually the summer thunderstorms which govern the sizing of the conveyance, treatment, and storage facilities in urban areas. Prolonged periods of snow melt can, however, produce very large quantities of runoff. But due to the slow rates of runoff, the only type of facilities it may affect are percolation and infiltration basins. These can take a long time to empty, and prolonged large inflows can overtax them. As a result, it is always wise to limit such facilities to the control of runoff from small tracts of land.

16.6 CHRONOLOGIC SERIES OF MEASURED FLOW DATA

The most direct method for sizing a detention storage facilities uses a chronologic series of actually measured flows at the storage site. Unfortunately, such data is very rare and, when available, is mostly limited to measurements at wastewater treatment plants. As a result, for practical reasons this technique is generally used for the design of storage at, or near, treatment plants in combined sewer systems.

Flow variations in a combined sewer can be categorized into the following groups:

- cyclic diurnal variations due to the daily rhythm of human activity;
- temporary shock loads due to releases from industry, swimming pools, water treatment plants, etc.;
- temporary shock loads due to discontinuities in operation of the upstream system (e.g., pumps, storage releases, etc.);
- temporary shock loads due to runoff from rainstorms;
- seasonal variations due to discontinuities in the use of water by schools, restaurants, hotels, industry, etc.; and
- cyclic variations in groundwater infiltration.

These types of variations in flow can act together and be additive. This concept is illustrated in Figure 16.12, where the runoff from a rainstorm is superimposed on the cyclic daily dry weather flow variations.

16.6.1 Daily Variations in Wastewater Flow

Dry weather flow of wastewater in a sewer has a characteristic diurnal variation which is dependent, among other things, on the size of the tributary area and its characteristics, such as the population density, industry, etc. As the tributary watershed increases, the variations in flow decrease. Conversely, the flow variations are most pronounced in sewers serving small tributary watersheds. Daily inflow can be intensified by temporary shock loads from industry, discontinuities in operation of pumping stations, etc.

Diurnal dry weather flow variations can be equalized with relatively moderate detention volumes. The EPA (1947) reported that flow equalization of the diurnal wastewater flow variations can be accomplished with a storage volume of 10% to 20% of the average one-day inflow. Of course, local conditions of flow, infiltration, etc. can significantly affect these storage requirements.

16.6.2 Shock Loads Due to Rainfall

In combined sewer systems, rainfall runoff significantly increases the flow in the sewers (see Figure 16.12). The magnitude of the flow increase is a function of many factors, including:

- type and age of the sewer system,
- layout of the system,
- intensity and duration of rainfall, and
- land uses in the tributary basin.

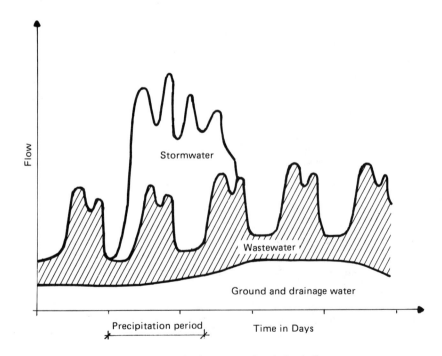

Figure 16.12   Various types of variation in flow.

Increases in flow in separate wastewater systems also occur during rain-storms due to inadvertent inflow. However, the increases in the peak flow are not nearly as pronounced as in combined sewer systems.

Large storage volumes are needed to provide flow equalization for rainfall-induced shock loads. For practical reasons, it is not possible to achieve complete flow equalization or to control the flows resulting from larger storms.

16.6.3 Variations Due To Snow Melt

Snow melt will also cause an increase in flow in combined sewers and storm sewers. The duration of snow melt can vary from a few hours to several weeks. The intensity and duration of snow melt within the sewer system will depend on the following:

- the climate,
- the amount of snow pack,
- the type of conveyance system, and
- temperature conditions.

The increase in flow within the sewers will continue after all the snow has melted. This is due to infiltration of the snow melt into the ground, which will in part seep into the sewer system.

Because snow melt can occur over a long period of time and produce large runoff volumes, it is unrealistic to build flow equalization storage for this runoff. Thus, in combined sewer systems, this flow increase is handled by providing additional hydraulic capacity at treatment plants.

16.6.4 Variations Due To Groundwater Seepage

Groundwater will seep into any sewer system as long as it in part, or totally, inundates any sewer pipe, manhole, inlet, etc. The infiltration rate will vary with the depth of groundwater and the condition and age of the sewer system. The older and more deteriorated systems have loser pipe joints, imperfect seals at manholes, and at other structures which act as conduits between the groundwater and the sewer pipes. Seepage rates fluctuate very slowly, and these slow variations cannot be effectively balanced through de-tention. It is suggested that measures be taken to reduce infiltration and exfiltration through restorative maintenance whenever it becomes a problem.

REFERENCES

ABRAHAM, C., LYONS, T. C., AND SCHULZE, K.W., "Selection of a Design Storm for Use with Simulation Models," *National Symposium on Urban Hydrology, Hydraulics and Sediment Control*, Lexington Ky., 1976.

ARNELL, V., "Rainfall Data for Design of Detention Basins," Seminar on Detention Basins in Sweden, 1978. (In Swedish)

ARNELL, V., "Rainfall Data for the Design of Sewer Pipe Systems," Chalmers University of Technology, Report Series A:8, Sweden, 1982.

BENGTSSON, L., JOHNSSON, A., MALMQUIST, P-A., SARNER, E., AND HALLGREN, J., "Snow in Urban Areas," Swedish Council for Building Research, Report R 27, 1980. (In Swedish)

EPA, *Flow Equalization,* EPA Technology Transfer Seminar Publication, 1974.

GEIGER, W., AND DORSCH, H., "Quality-quantity Simulation (QQS), Detailed Continuous Planning Model for Urban Runoff Control," Volume I, EPA Grant No. R 805100, 1980.

HUFF, F. A., "Time Distribution of Heavy Rainfall Storms," *Water Resources Research,* Vol. 3, No. 4, 1967.

KEIFER, C. J., AND HSIEN CHU, H., "Synthetic Storm Pattern for Drainage Design," *Journal of the Hydraulics Division,* Vol. 83, ASCE, August, 1957.

MARSALEK, J., "Research on the Design Storm Concept," *ASCE Urban Water Resources Research Program, Technical Memorandum No. 33,* ASCE, New York, 1978.

MARSALEK, J., URBONAS, B., ROSSMILLER, R., AND WENZEL, H., (i.e., UWRRC Design Storm Task Committee), "Design Storms for Urban Drainage," *Proceedings, Water Forum '86,* ASCE, August, 1986.

MCPHERSON, M. B., "Urban Hydrology: New Concepts in Hydrology for Urban Areas," Northwest Bridge Engineering Seminar, Olympia, Wash., October, 1976.

MCPHERSON, M. B., "The Design Storm Concept," *Urban Runoff Control Planning,* ASCE Urban Water Resources Research Council, ASCE, June, 1977.

SIFALDA, V., "Development of a Design Rain for Assigning Dimensions to Sewer Nets," Gwf, Wasser/Abwasser No. 9, 1973. (In German)

TERSTRIEP, M. B., AND STALL, J. B., *The Illinois Urban Drainage Area Simulator, ILLUDAS,* Bulletin 58, Illinois Water Survey, Urbana, 1974.

THORNDAL, U., "Precipitation Hydrographs," Stads- og Havneingsnioren No. 7, Copenhagen, 1971. (In Danish)

URBAN STORM DRAINAGE CRITERIA MANUAL, Urban Drainage and Flood Control District, Denver, Colo., 1984 edition.

URBONAS, B., "Reliability of Design Storms in Modeling," *Proceedings of the International Symposium on Urban Storm Runoff,* University of Kentucky, July, 1979.

WALESH, S. G., "Statistically-Based Use of Event Models," *Proceedings of the International Symposium on Urban Storm Runoff,* University of Kentucky, July, 1979.

WALESH, S. G., "Summary—Seminar on the Design Storm Concept," *Proceedings, Stormwater Management Model (SWMM) Users Group Meeting,* May, 1979, EPA 600/9-79-026, June, 1979.

WENZEL, H. G., JR., "Rainfall Data for Sewer Design," *Section III of Storm Sewer Design,* Dept. of Civil Engineering, University of Illinois-Urbana-Champaign, 1978.

17

Calculating Infiltration and Percolation Facilities

17.1 GENERAL

The design of infiltration and percolation facilities is similar to the design of other types of detention facilities. The goal is to design the facility so that it will contain the design inflow without overflowing. At infiltration facilities, water is infiltrated into the ground from the surface. If the flow to the infiltration surface exceeds its infiltrating capacity, excess water is stored on the surface. At percolation facilities, the water is conveyed to a percolation pit from where it percolates out into the ground. In a percolation basin, the effective pore volume serves as the storage volume.

As a general note, infiltration and percolation facilities will work best for very small runoff basins such as individual lots. It cannot be overemphasized that they need to be designed conservatively, using low hydraulic unit loading rates on the infiltration/percolation surfaces. These facilities need to "rest" between runoff events to "heal" and rejuvenate the pores in the receiving soils. If the loading rates do not permit this, the soil's pores seal and the infiltration capacity of the site may be lost, possibly forever.

The installation and use of infiltration and percolation facilities was described in Chapter 2, which introduced the concept of local disposal facilities, explained some of the different concepts in arranging such facilities, and discussed their suitability under various soil and groundwater conditions. It is suggested that the reader review Chapter 2 before proceeding to use the principles described in the remainder of this chapter.

17.2 DESIGN FLOW

The volume of runoff reaching an infiltration or a percolation facility depends on several factors. These include the tributary basin size, the degree of development in the basin (i.e., amount of impervious surface), and the characteristics of rainfall and snow melt in the area. We begin by discussing the suggested techniques for estimating rainfall and snow melt runoff for the design of infiltration and percolation facilities.

17.2.1 Stormwater Runoff

Since infiltration and percolation facilities are mainly used for small runoff basins, runoff calculations can be based on the Rational Formula; thus, in Metric units:

$$Q_T = C \cdot \frac{I_T}{1,000} \cdot A \qquad (17.1)$$

in which Q_T = runoff rate for a T-year storm, cubic meters/second,
$\quad\quad\;\; C$ = runoff coefficient, nondimensional,
$\quad\quad\;\; I_T$ = rainfall intensity for a T-year storm at a storm duration t, in liters/second/hectare, and
$\quad\quad\;\; A$ = area of the tributary watershed, in hectares.

By multiplying the average runoff rate, Q_T, by the design storm duration, t, we obtain the cumulative volume over the storm duration, namely

$$V_T = 3,600 \cdot C \cdot \frac{I_T}{1,000} \cdot t \cdot A \qquad (17.2)$$

in which V_T = total runoff volume at time t for a T-year storm in cubic meters, and
$\quad\quad\;\; t$ = storm duration in hours.

Thus, the volume calculations can be performed using the intensity–duration–frequency curves for any design storm having a T-year return period. By deciding on the design storm duration, the volume of rainfall (i.e., block rain = $I_T t$) can be calculated, the simplification being justified for small urban watersheds. This procedure is illustrated in Figure 17.1 for three storm durations.

Block rain represents only the average intensity of the most intense portion of the rainstorm. The rain that falls before and after this period is not

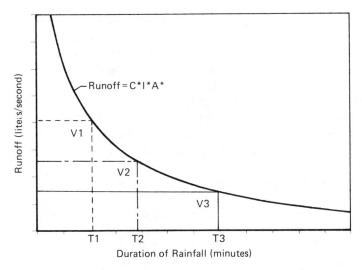

Figure 17.1   Derivation of block rain inflow hydrographs.

included in the I–D–F curve and consequently is not reflected in the block rain calculations. Since the sizing of an infiltration or a percolation basin depends mostly on volume of runoff, it is necessary to somehow account for that part of the rainstorm not included in block rain.

Sjoberg and Martensson (1982) studied how the runoff from block rainfall differed from the results obtained using continuous simulation with chronological rainfall records. Assuming the latter provided more accurate runoff estimates, they concluded that by increasing block rainfall volume by 25%, the runoff volume estimates with Rational Formula can be quite accurate. This modification does not affect the simplicity of the calculations. Thus, the volume of runoff reaching an infiltration or a percolation facility can be estimated using Equation 17.3, which is a slightly modified version of Equation 17.2:

$$V_T = 1.25 \left(3{,}600 \cdot C \cdot \frac{I_T}{1{,}000} \cdot t \cdot A \right) \qquad (17.3)$$

in which all of the terms have been defined earlier.

When sizing infiltration and percolation facilities, it is practical to assume that all runoff occurs only from the impervious surfaces having runoff coefficients between 0.85 and 0.95. It is not uncommon to design infiltration and percolation facilities for the average runoff volume from all the storms that occur over a period of years. The average runoff event is considerably smaller than the runoff from a two-year design storm. This is because the storm data include *all* runoff-producing events and not only the most intense storms of each year, which is the case for typical I–D–F curves. For example, in the Denver area the average runoff producing storm has 0.44 inches of precipitation. The two-year design storm, on the other hand, has approximately 1.10 inches of rainfall. Because the average runoff events are relatively small, facilities sized to handle them will be loaded to full capacity much more frequently. As a result, they will be stressed and the infiltration and percolation surfaces may not have a chance to sufficiently rest between storms.

The choice of the recurrence frequency and storm duration is often dictated by local or state government rules, regulations, or criteria. Nevertheless, for most stormwater drainage disposal situations it is recommended that the design for these type of facilities be based on at least a two-year storm. This will insure a longer life for the infiltration and percolation facilities.

17.2.2 Snow Melt

In certain parts of the United States and Europe, snow melt can govern the sizing of infiltration and percolation facilities. This is especially the case when the basins have a relatively small amount of impervious surface compared to the previous surface. Under such conditions, the runoff from snow melt may produce prolonged and larger volumes of runoff than rainfall.

Under extreme conditions, actual snow melt intensities could reach as much as 0.15 inches of water per hour. However, more typical runoff rates from melting snow appear to be considerably less than this rate.

Unlike rainfall on unfrozen and unsaturated soils, snow melt will contribute to surface runoff from the pervious surfaces. However, it is not possible to generalize how much runoff can be expected under the varying climatic conditions throughout the United States and Europe. To insure acceptable operation, check the design of an infiltration or percolation facility against the runoff from melting snow. We suggest that this be done using the following minimum snow melt rates:

Minimum Snow Melt Rates for Design

IMPERVIOUS SURFACES		PERVIOUS SURFACES	
English Units	Metric Units	English Units	Metric Units
0.04 ft³/s/ac	2.8 l/s/ha	0.02 ft³/s/ac	1.4 l/s/ha

17.3 SELECTING AND SIZING INFILTRATION SURFACES

17.3.1 Site Selection for Infiltration

Many factors affect the suitability of a site as an infiltration facility for the disposal of stormwater. Among these, the following are most important:

- depth to groundwater;
- depth to bedrock;

- surface soil type;
- underlying soil type;
- vegetation cover of the infiltrating surface;
- the uses of the infiltrating surfaces; and
- the ratio of tributary impervious surface to the infiltrating surface.

There are several conditions that will rule out a site as a candidate for an infiltration facility. **If the following conditions are discovered or likely, disposal of stormwater by infiltration is not recommended:**

- seasonal high groundwater is less than 4 feet below the infiltrating surface.
- bedrock is within 4 feet of the infiltrating surface.
- the infiltrating surface is on fill (unless the fill is clean sand or gravel).
- the surface and underlaying soils are classified by the Soil Conservation Service as Hydrologic Group D, or the saturated infiltration rate is less than 0.3 inches per hour as reported by SCS soil surveys.

If the preceding conditions do not rule out the site, the site should be evaluated using a method developed by the Swedish Association for Water and Sewer Works (1983). This procedure is based on evaluating a series of site conditions and assigning points for each one of them. If a site gets less than 20 points, it should not be considered for infiltration. On the other hand, a site with more than 30 points is considered to be excellent. Sites receiving 20 to 30 points are considered good candidates for an infiltration facility. Figure 17.2 was prepared to summarize this process and to assist the reader in evaluating infiltration sites.

The points to be assigned for each of the site characteristics are tabulated in Table 17.1. Points are assigned for each site condition and are then added to determine the total points for the site. This evaluation system should only be used for preliminary screening of potential sites. It is not intended as a substitute for good engineering and site-specific testing, evaluation, and design.

17.3.2 Sizing an Infiltration Facility

When the site has been judged to be an acceptable candidate for an infiltration basin, the next step is to find the required surface area and storage volume for the facility. Table 17.1 indicates that the infiltration surface area should not be smaller than one-half of the tributary impervious surface. This requirement sets the lower limit for the size of the infiltration surface. There is no upper limit, and the size of the infiltration surface will be constrained only by the available right-of-way.

The size of the infiltration surface will also be governed by the release rate, namely the infiltration rate for the water stored on the infiltration sur-

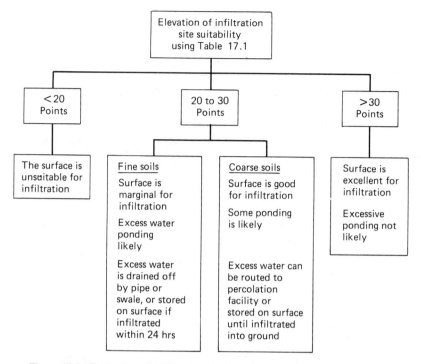

Figure 17.2 Evaluation of infiltration site for suitability. (After Swedish Association for Water and Sewage Works, 1983.)

face. The exact value of the infiltration rate is one of the most difficult parameters to determine. Among the factors influencing infiltration rates are:

- the structure of the ground surface,
- the type and condition of the vegetation zone,
- soil moisture,
- the nature of the underlying soils, including permeability, and
- the depth to groundwater.

It is not possible to generalize about infiltration rates. Table 17.2 contains data that illustrate their diversity among various soil groups. Clearly, the only practical recommendation we can make is to perform several infiltration tests at each site and then to use the lowest measured rates in the design of an infiltration facility.

When the inflow envelope curve (see Equation 17.3) and the infiltration rate are known, the storage volume can be calculated graphically. As illustrated in Figure 17.3, the required storage volume is found by measuring the largest vertical difference between the inflow envelope function, $V(t)$, and the cumulative infiltration function $F(t)$. Note that this graphical procedure can easily be programmed on any commonly available spreadsheet.

TABLE 17.1   Point System for Evaluating Infiltration Sites

---

1. Ratio between tributary connected impervious area (A_{IMP}) and the infiltration area (A_{INF}):
   - $A_{INF} > 2\,A_{IMP}$ 20 points
   - $A_{IMP} \leq A_{INF} \leq 2\,A_{IMP}$ 10 points
   - $0.5\,A_{IMP} \leq A_{INF} < A_{IMP}$ 5 points

   Pervious surfaces smaller than $0.5\,A_{IMP}$ should not be used for infiltration.

2. Nature of surface soil layer:
   - Coarse soils with low ratio of organic material   7 points
   - Normal humus soil   5 points
   - Fine-grained soils with high ratio of organic material   0 points

3. Underlaying soils:
   - If the underlaying soils are coarser than surface soil, assign the same
     number of points as for the surface soil layer assigned under item 2.
   - If the underlaying soils are finer grained than the surface soils,
     use the following points:
     - Gravel, sand, or glacial till with gravel or sand   7 points
     - Silty sand or loam   5 points
     - Fine silt or clay   0 points

4. Slope (S) of the infiltration surface:
   - $S < 7\%$ 5 points
   - $7 \leq S \leq 20\%$ 3 points
   - $S > 20\%$ 0 points

5. Vegetation cover:
   - Healthy, natural vegetation cover   5 points
   - Lawn—well-established   3 points
   - Lawn—new   0 points
   - No vegetation—bare ground   −5 points

6. Degree of traffic on infiltration surface:
   - Little foot traffic   5 points
   - Average foot traffic (park, lawn)   3 points
   - Much foot traffic (playing fields)   0 points

TABLE 17.2   Typical Infiltration Rates

SCS Group and Type	Infiltration Rate (Inches per Hour)
A. Sand	8.0
A. Loamy sand	2.0
B. Sandy loam	1.0
B. Loam	0.5
C. Silt loam	0.25*
C. Sandy clay loam	0.15
D. Clay loam and silty clay loam	<0.09
D. Clays	<0.05

* Minimum rate; soils with lesser rates should not be considered as
candidates for infiltration facilities.

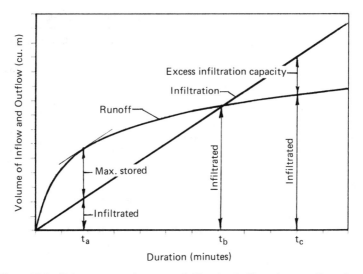

Figure 17.3   Sizing storage volume at an infiltration facility using runoff envelope.

This figure has two distinct regions. For storms having runoff duration equal to or less than t_b, all of the runoff cannot infiltrate into the ground. When the duration of rain is greater than t_b, infiltration capacity exceeds the inflow volume and the stored water is infiltrated into the ground.

These hydrologic calculations tell us how much runoff is not infiltrated into the ground for a given I–D–F condition. This excess either runs off or is stored on the infiltration surface. As an alternative, the infiltration area can be increased so that the runoff envelope curve $V(t)$ never exceeds the cumulative infiltration line $F(t)$. The designer may wish to try several infiltration area sizes until a satisfactory balance is achieved between storage volume and the size of the infiltration surface. Do not hesitate to oversize the area if more grassed surface area is available at the site than is required for a minimum design. Take full advantage of the site to reduce loading rates on the infiltration surfaces, thereby increasing longevity and reducing the period of wet site conditions.

17.4 SELECTING AND SIZING PERCOLATION BASINS

17.4.1 Site Selection for Percolation Basins

The factors affecting the suitability of a site for percolation are similar to those affecting infiltration facilities, namely:

● depth to groundwater;
● depth to bedrock;

- soil type adjacent to and below the percolation bed; and
- ratio of tributary impervious surface to the percolation surface area.

As with infiltration, there are site conditions that can rule out a site as a viable candidate. **If the following conditions are discovered or are likely occur at the site, disposal of stormwater by percolation is not recommended:**

- Seasonal high groundwater is less than 4 feet below the bottom of percolation bed.
- Bedrock is within 4 feet of the bottom of the percolation bed.
- Percolation bed is located on fill unless the fill is clean sand of gravel.
- The adjacent and underlying soils are classified by the Soil Conservation Service as Hydrologic Group C or D, or the field tested saturated hydraulic conductivity of the soils is less than 2×10^{-5} meters per second.

When these do not rule out the site, we again suggest that the percolation facility be designed using a method developed by the Swedish Association for Water and Sewer Works (1983). Since this procedure comes from Europe, all equations are in metric units.

17.4.2 Darcy's Law

The rate at which water percolates into the ground can be estimated with the aid of Darcy's Law, namely:

$$U = k \cdot I \qquad (17.4)$$

in which U = flow velocity in meters per second,
$\quad k$ = hydraulic conductivity in meters per second, and
$\quad I$ = hydraulic gradient in meters per meter.

To be precise, Darcy's Law applies to groundwater flow in saturated soils. However, we try to locate percolation facilities with their bottoms at least 4 feet above the seasonal high ground-water table. It is nevertheless safe to assume that the soil will be saturated when the facility is operating and that, because the bottom of the percolation field is above normal groundwater, the hydraulic gradient $k = 1.0$ meters/meter.

It is not possible to generalize what hydraulic conductivity should be used and we recommend that percolation tests be performed at each individual site. Table 17.3 contains published ranges of hydraulic conductivity for various types of soils. As you can see, the value of conductivity for any soil type can vary by as much as four orders of magnitude, which further reinforces the need for site-specific data.

When performing field hydraulic conductivity tests, unless there is good reason, the lowest conductivity test value is the one that should be used.

Remember that the soils will tend to gradually clog with time, and the available conductivity will decrease. For this reason, it is recommended that the field conductivity test values are reduced by a safety factor of 2 to 3 when designing any percolation facility. Keep in mind that once a percolation facility fails, it will be very expensive or impossible to rebuild. Therefore, a conservative design philosophy is recommended. Designing a system that has a chance of failing in a few years is considered to be poor engineering practice.

17.4.3 Effective Porosity of Percolation Media

The effective porosity of the porous fill media inside of the percolation pit of trench determines the volume that is available for storage of water. Table 17.4 lists representable values for some of the most commonly used materials.

17.4.4 Effective Percolation Area

It is recommended by the Swedish Association for Water and Sewer Works that the bottom of a percolation pit or trench is considered impervious. The reason for this is that the bottom seals quickly by the accumulation of sediments. It is a good idea to anticipate this by assuming that all water will percolate into the ground only through the vertical sides of the basin or trench.

TABLE 17.3   Hydraulic Conductivity of Several Soil Types

Soil Type	Hydraulic Conductivity (Meters per Second)
Gravel	10^{-3}–10^{-1}
Sand	10^{-5}–10^{-2}
Silt	10^{-9}–10^{-5}
Clay (saturated)	$<10^{-9}$
Till	10^{-10}–10^{-6}

TABLE 17.4   Effective Porosity of Typical Stone Materials

Material	Effective Porosity (Percent)
Blasted rock	30
Uniform sized gravel	40
Graded gravel (¾-inch minus)	30
Sand	25
Pit run gravel	15–25

17.4.5 Calculating the Needed Storage Volume

The outflow rate from a percolation facility can be estimated using Darcy's Law. The water depth in the basin will vary during the filling and emptying process. To approximate the average release rate, the water depth can be arbitrarily set at one-half of the maximum depth. This means that the effective percolation area will be equal to one-half the area of the sides of the basin. With a hydraulic gradient equal to 1.0, Darcy's Law gives the following expression for the outflow from the basin:

$$V_{out}(t) = k \cdot 1.0 \cdot \frac{A_{perc}}{2} \cdot 3{,}600 \cdot t \qquad (17.5)$$

in which $V_{out}(t)$ = volume of water percolated into the ground, in cubic meters,

$\quad\quad$ k = hydraulic conductivity of soil, in meters per second,

$\quad\quad$ A_{perc} = total area of the sides of the percolation facility, in square meters, and

$\quad\quad$ t = percolation time, in hours.

The volume of water stored, V, in the facility is the maximum difference between $V_{in}(t)$ and $V_{out}(t)$, which can be expressed as follows:

$$V = \max[V_{in}(t) - V_{out}(t)] \qquad (17.6)$$

or

$$V = \max\left[t \cdot 3{,}600 \cdot \frac{I_t}{1{,}000} \cdot C \cdot A \cdot 1.25 - k \cdot \frac{A_{perc}}{2} \cdot 3{,}600 \cdot t\right] \qquad (17.7)$$

in which A = area of the tributary impervious surfaces in hectares,

$\quad\quad$ C = runoff coefficient for the impervious surfaces, and

$\quad\quad$ I_t = average rainfall intensity over time t, in liters/second/hectare.

If both sides of Equation 17.7 are divided by $(C \cdot A)$, we get,

$$\frac{V}{C \cdot A} = \max\left[1.25 \cdot t \cdot 3.6 \cdot I_t - k \cdot \frac{A_{perc}}{2} \cdot 3{,}600 \cdot \frac{t}{(C \cdot A)}\right] \qquad (17.8)$$

and if in Equation 17.8 the following substitutions are made:

$$D = \frac{V}{C \cdot A} \qquad (17.9)$$

$$E = \frac{1{,}000 \cdot k \cdot \dfrac{A_{perc}}{2}}{C \cdot A} \qquad (17.10)$$

we get:

$$D = \max[4.5 \cdot I_t \cdot t - 3.6 \cdot E \cdot t] \qquad (17.11)$$

In this expression, the parameter D represents the specific percolation volume, which is the storage volume expressed in cubic meters per hectare of connected impervious area. The parameter E represents the specific outflow from the basin expressed as liters per second and hectare of connected impervious area.

As a first step, an envelope curve of the specific inflow, V_{in}, has to be constructed. This is done by graphing the expression $(4.5 \cdot I_t \cdot t)$ for different values of t. The values of I_t are taken from the intensity–duration–frequency (I–D–F) curve used for the region or location. An example of this is illustrated in Figure 17.4.

Various specific outflow rates, V_{out}, can now be superimposed on the same diagram as straight lines. The specific storage volumes, D, for different specific outflow rates can be obtained graphically as the maximum vertical distance between the envelope curve and the outflow line (see Figure 17.5). The corresponding values of D are plotted against the values of E, as shown in Figure 17.6. Figure 17.6 shows a curve for a single recurrence design rainstorm frequency. If desired, this same process can now be repeated for other recurrence rainfall frequencies. The actual sizing of a percolation facility can now be performed using a simple trial-and-error procedure described next.

17.4.6 Step-by-Step Calculating Procedure

In practice, percolation facilities can be dimensioned using the following step-by-step procedure:

1. Construct a design curve similar to Figure 17.6, where the specific storage volume D is plotted as a function of the specific outlet capacity E.
2. Estimate the area of the impervious surfaces hydrologically connected to the percolation facility.

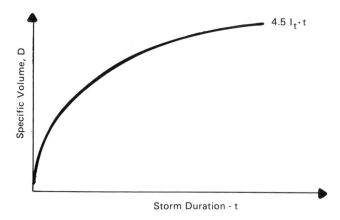

Figure 17.4   Envelope curve of specific inflow V_{in} to a percolation facility.

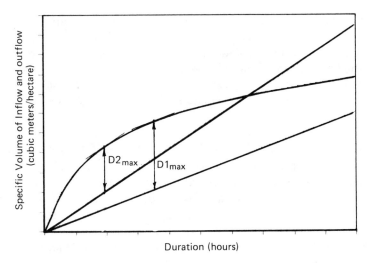

Figure 17.5 Graphic calculation of D, the specific storage.

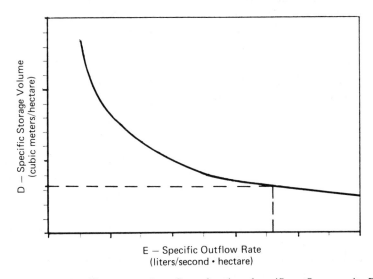

Figure 17.6 Specific storage volume D as a function of specific outflow capacity E.

3. Calculate the specific outflow capacity E for the percolation facility in question. This is done using Equation 17.10, namely,

$$E = \frac{1{,}000 \cdot k \cdot \dfrac{A_{perc}}{2}}{C \cdot A}$$

In this expression, the value of A the impervious area, C its runoff coefficient, and k the hydraulic conductivity of the soil are known. As we suggested earlier, k values measured in the field should be first multiplied by a factor of 0.3 to 0.5 as a safety factor. Thus, after the k value is determined, one has to estimate the total area of the sides of a percolation facility (A_{perc}) to calculate E. This area is expressed as

$$A_{perc} = 2 \cdot (a + b) \cdot h \qquad (17.12)$$

in which a = length, b = width, and h = height of the stone filled trench. The value of h must be chosen after considering local conditions and keeping in mind that the entire percolation facility must lie above the highest seasonal groundwater level. Since the outflow is expected to take place only through the sides, the design should strive to maximize the area of the sides and not the bottom. The practical consequence of this is that most efficient percolation facilities are relatively long and narrow. As a practical matter, the width b and the height h are normally set between one and two meters.

4. When the value of the parameter E has been calculated, the specific storage volume D is found on the D vs. E curve (see Figure 17.6).

5. The necessary storage volume is calculated using:

$$V_{storage} = D \cdot (A \cdot C) \qquad (17.13)$$

6. If the effective porosity of the stone filling in the trench is n, the total volume of the basin will be:

$$V_{basin} = \frac{V_{storage}}{n} \qquad (17.14)$$

7. The calculated total basin volume, V_{basin}, is then compared to the volume found using the assumed $(a \cdot b \cdot h)$ dimensions.

    If V_{basin} is less than $(a \cdot b \cdot h)$, the assumed basin was undersized and the calculations are repeated with larger values of a, b, or h. If V_{basin} is greater than $(a \cdot b \cdot h)$, then the assumed basin is oversized and the calculations can be repeated with smaller values for a, b, or h until a reasonable agreement is achieved between the assumed basin volume and the one calculated as required to do the job.

17.4.7 Checking Snow Melt Condition

All percolation facilities should be checked to be sure that they will not be overloaded by melting snow. During prolonged snow melt periods, the facility may operate near full depth. Because of the prolonged runoff period, the water depth can be assumed to be at full height of the trench or basin.

Thus, the area of the percolation facility, namely the area of the sides of the trench or basin, under snow melt conditions may be calculated using:

$$A_{perc} = 2 \cdot (a + b) \cdot h \tag{17.15}$$

The designer must now check to see if the percolation facility sized to dispose of rainfall runoff can handle the runoff from melting snow. Unlike rainfall, runoff from melting snow is expected to occur from both the impervious and pervious surfaces. As a result, the following condition has to be satisfied:

$$(A_{perc} \cdot k \cdot I \cdot 1,000) > (S_{perv} \cdot A_{perv} + S_{imperv} \cdot A_{imperv}) \tag{17.16}$$

in which S_{perv} = design snow melt runoff rate from the pervious surfaces, and S_{imperv} = design snow melt runoff rate from the impervious surfaces.

The preceding expression can also be written as:

$$[h \cdot 2 \cdot (a + b) \cdot k \cdot 1,000] > (S_{perv} \cdot A_{perv} + S_{imperv} \cdot A_{imperv}) \tag{17.17}$$

When the snow melt rates from pervious and impervious surfaces are the equal, the preceding equation becomes:

$$[h \cdot 2 \cdot (a + b) \cdot k \cdot 1,000] > S \cdot (A_{perv} + A_{imperv}) \tag{17.18}$$

17.4.8 Auxiliary Outlet Pipe

Sometimes, even when all of the selection criteria are met, it may still not be practical to construct the required storage volume for a percolation facility. This can occur when the soils do not have sufficient hydraulic conductivity to empty the facility in a reasonable amount of time using only percolation. In some of these situations, it may be possible to install an auxiliary outlet pipe. This pipe collects the water from the percolation basin and releases it slowly through an orifice or a flow throttling valve. The following rule-of-thumb guidelines can be used to evaluate the need for an auxiliary pipe:

1. Whenever the soils have a hydraulic conductivity that is less than 2×10^{-5} meters per second, the percolation rate is considered zero and the site declared unsuitable.

    If, however, a percolation facility is installed despite this recommendation, it should be designed assuming that the entire outflow will occur through the auxiliary outlet pipe. Normally, such an installation is unjustified, and we suggest that the designer investigate an alternative type of a detention facility, such as pipe packages, underground tanks, surface basins, etc.

2. When soils have hydraulic conductivity greater than 5×10^{-4} meters per

second, it is safe to assume that all emptying takes place through percolation into surrounding soils.

3. An auxiliary outlet pipe may be feasible in soils with hydraulic conductivity between 2×10^{-5} and 5×10^{-4} meters per second. Such a pipe can provide a margin of safety and insure that the facility drains in a reasonable amount of time.

4. The auxiliary outlet pipe should always be equipped with a flow restrictor. This restrictor is designed to provide a total outlet rate (i.e., percolation through soil and auxiliary outlet) equivalent to a basin having a percolation rate of 5×10^{-4} meters per second.

If, however, the percolation basin is being provided primarily to improve stormwater runoff quality, the emptying pipe would not produce the desired results. With water quality, the goal is to trap and infiltrate into the ground as much of the runoff as possible. As a result, the auxiliary emptying pipe should be located near the top of the percolation pit and act as an uncontrolled emergency overflow. The water trapped below this pipe may eventually percolate into the ground and receives filtering as it flows through the soils. Such a system for disposal of runoff to improve water quality should never be used when soil permeability k is less than 4×10^{-5} meters per second, and should be seriously questioned whenever k is less than 10^{-4} meters per second.

17.5 DESIGN EXAMPLES

17.5.1 Evaluation of an Infiltration Site

An infiltration site is connected to a 100 square meters roof. The infiltration surface is a 210 square meter lawn with a slope of 10%. The topsoil and the underlaying soils are composed mostly of coarse silt. Check to see if the lawn is a good candidate for infiltration.

The site is first evaluated using the point system listed in Table 17.1. The results are as follows:

1. The ratio between the impervious surface and the infiltration surface areas is

$$A_{inf} = 2.1 \, A_{imp}$$

This gives the site *20 points*.

2. The topsoil is coarse silt. This gives the site *5 points*.

3. The underlaying soil is coarse silt. This gives the site *5 points*.

4. The slope of the infiltration surface is 0.10 ft/ft. This gives the site *3 points*.

5. The infiltration surface is a new established lawn. This gives the site *0 points*.

6. It is expected the lawn will have normal foot traffic. This gives the site *3 points.*

7. The total is *36 points* for this site evaluation. According to the guidelines shown in Figure 17.2, the site can be used for infiltration. Surface runoff from this site is not likely to occur, except during larger than usual storms. Now proceed to size the needed storage volume above the infiltration surface using the runoff envelope method.

17.5.2 Sizing of a Percolation Facility

In this example, runoff from a roof with an area of 800 square meters (assume the runoff coefficient C = 1.0) and a lawn of another 800 square meters is led to a percolation facility. The design parameters and other constraints are listed following. Find the length of the percolation trench:

Tributary area:

$$A_{imp} = 800 \text{ m}^2 \qquad A_{perv} = 800 \text{ m}^2$$

Percolation facility:

Maximum height of stone filled trench = 0.8 m.
Maximum width of stone filled trench = 1.0 m.
Porosity of stone filling: $n = 0.4$
Minimum depth to seasonal groundwater = 2 m.
Hydraulic conductivity of soil: $k = 4 \times 10^{-5}$ m/s.

Design rainfall:
Two-year storm.

The calculation procedure is as follows:

1. Construct a design curve as shown in Figure 17.7 using the I–D–F curve for the two-year storm.

2. The native soils have a hydraulic conductivity of 4×10^{-5} m/s. To compensate for the uncertainties of the soils investigation, a safety factor is applied to this value. This assures that the percolation rate is not overstated and the facility is not underdesigned. Thus, the design hydraulic conductivity used in this example is $0.5 \times 4 \times 10^{-5} = 2 \times 10^{-5}$ meters per second.

3. For the first trial, assume the length of the percolation trench is 35 meters. The value for E is calculated using Equation 17.10:

$$E = \left[2 \cdot \left(35 + 1.0 \right) \cdot \frac{0.8}{20} \right] \cdot 2 \times 10^{-5} \cdot \frac{1,000}{0.08}$$

$$= 7.2 \text{ liters per second hectare}$$

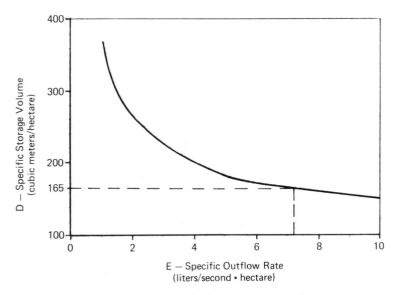

Figure 17.7 Design curve for example facility using a two-year design storm I–D–F.

4. Using Figure 17.7, we see that when $E = 7.2$, $D = 165$ cubic meters per hectare.

5. Since the impervious area tributary to the percolation trench is 0.08 hectares, the needed storage volume is found using Equation 17.11:

$$V_{storage} = 0.08 \cdot 165 = 13.2 \text{ cubic meters}$$

6. Given that the effective porosity of the stone fill media is 0.4, the total trench or basin volume is found using Equation 17.14:

$$V_{basin} = \frac{13.2}{0.4} = 33 \ m^3$$

7. The assumed total trench volume in step 3 was $V_{basin\text{-}assumed} = (35 \cdot 1.0 \cdot 0.8) = 28$ cubic meters, which is less than the calculated volume of 33 cubic meters.

We see that the assumed length of the trench does not provide sufficient basin volume. Therefore, the calculations are repeated starting with step 3 assuming a trench length of 40 meters. This time we get:

Calculated:

$$V_{basin} = 30 \text{ cubic meters}$$

vs.

Assumed:

$$V_{basin} = 32 \text{ cubic meters}$$

The second assumption resulted in slightly more volume than needed. However, this oversizing is relatively small, and the 40-meter-long trench is chosen for final design.

As a final step, the design is tested to see if it is adequate under snow melt conditions. Using the design snow melt rates recommended in section 17.2.2, check to see if the limit of Equation 17.12 is satisfied, namely:

$$A_{perc} \cdot k \cdot 1 \div 1,000 = 2 \cdot 0.8 \cdot (40 + 1.0) \cdot 2 \times 10^{-5} \cdot 1,000 = 1.312$$

$$S_{perv} \cdot A_{perv} + S_{imperv} \cdot A_{imperv} = 1.4 \cdot 0.08 + 2.8 \cdot 0.08 = 0.336$$

Since $1.312 > 0.336$, the snow melt condition does not govern, and the original design is considered acceptable for final design.

REFERENCES

SJOBERG, A., AND MARTENSSON, N. "Analysis of the Envelope Method for Dimensioning Percolation Facilities," Chalmers University of Technology, 1982. (In Swedish)

SWEDISH WATER AND SEWAGE WORKS ASSOCIATION, *Local Disposal of Storm Water,* Publication VAV P46, 1983. (In Swedish)

18

Calculation Methods for Detention Facilities

18.1 GENERAL

As the name implies, a detention facility temporarily detains stormwater runoff. They are used primarily to reduce the peak rate of flow in the downstream conveyance system so that its capacity is not exceeded. Sometimes, and more particularly now than in the past, detention is used to remove sediments in the runoff and, as a result, improve water quality. Typically, detention facilities are designed to have sufficient volume to control the peak rate of flow during a given design storm.

Detention storage volume is estimated by calculating the differences between the inflow and outflow hydrographs as illustrated in Figure 18.1. The basic equation for these calculations is

$$V = \int_{o}^{t_o} (Q_{in} - Q_{out}) \, dt \tag{18.1}$$

in which V = required stage volume,

t = time from beginning of storage,

t_o = time when the outflow hydrograph intersects the recession limb of the inflow hydrograph,

Q_{in} = inflow rate, and

Q_{out} = outflow rate.

The design inflow into a detention facility is usually defined by the design storm which is used to calculate an inflow hydrograph. If however, the storage basin experiences a hydrograph that is larger than it was designed to store, the excess water spills over the embankment or a spillway. The idealized shapes of the inflow and the outflow hydrographs under such a scenario are illustrated in Figure 18.2.

The volume of a detention storage basin can be calculated in many different ways. The following are some of the categories of methods that are used to calculate detention storage volumes:

- calculations without considering time of concentration,
- calculations considering time of concentration,
- time–area method,
- rain point diagram method,
- summation curve method, and
- detailed flow routing methods.

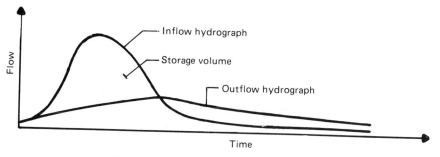

Figure 18.1   Determination of storage volume.

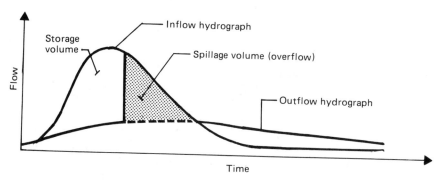

Figure 18.2   Hydrographs when the storage capacity is exceeded.

18.2 CALCULATIONS WITHOUT CONSIDERING
TIME OF CONCENTRATION

As described earlier, calculations of runoff using the Rational Formula are based on block rain having a uniform rainfall intensity for the duration of the rainstorm. Information on the intensity of the rain is obtained from intensity–duration–frequency (i.e., I–D–F) curves. Kropf and Geiser (1957), FAA (1966), DRCOG (1961), and others have described a simplified method for the sizing of detention storage using I–D–F curves and the Rational Formula. The technique does not require the use of the time of concentration for the tributary watershed.

As a first step in this procedure, it is necessary to calculate the total runoff volume for a range of storm durations. This is illustrated in Figure 18.3. The inflow volume for any storm duration is merely the product of time (i.e., duration) and the runoff rate for that duration (i.e., time of concentration), namely,

$$V = T \cdot C \cdot I \cdot A \tag{18.2}$$

in which V = runoff volume, in cubic feet,
$\quad T$ = storm duration, in seconds,
$\quad C$ = runoff coefficient,
$\quad I$ = average storm intensity for storm duration T,
$\qquad$ in inches/hour, and
$\quad A$ = area of the watershed, in acres.

The next step is to construct a runoff envelope curve similar to the one described in Chapter 17. This is done by plotting the runoff volumes calculated for each duration, as shown in Figure 18.4. To complete the calculations, superimpose on the runoff envelope diagram the line representing the desired release rate from the storage basin. The required volume is then obtained by

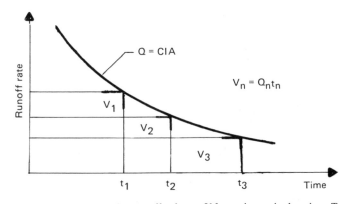

Figure 18.3   Calculating runoff volumes V for various rain durations T.

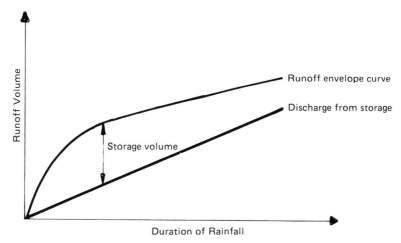

Figure 18.4   Finding the storage volume graphically using the runoff envelope diagram.

finding the maximum difference between the runoff envelope and the release discharge line. This process can be tabularized using any commercial spread sheet program, and the plotting of the runoff envelope diagram can be by-passed.

The preceding example was for a single recurrence frequency storm. If one wants to calculate volumes for other frequencies, the calculations have to be repeated and a runoff envelope diagram constructed for each design frequency.

According to the DRCOG (1961), this procedure tends to overestimate the runoff and storage volume for larger urban watersheds. They recommend that the Rational Method and all procedures related to it be limited to water-sheds having an area that is less than 160 acres.

18.3 CALCULATIONS CONSIDERING TIME OF CONCENTRATION

The aforementioned method calculates the storage volume without consider-ing the time of concentration of the watershed. Now we describe a technique that uses the Rational Method and considers the time of concentration. This technique and its many variations were described by Lautrich (1956), Annen and Londong (1960), Malpricht (1962), Pecher (1970), Kao (1975), and APWA (1974, 1981).

The required storage volume is determined by finding the maximum difference between the areas under the trapezoidal inflow hydrographs and the desired basin release discharge rate K. This procedure is illustrated graph-ically in Figure 18.5. When the storm duration is equal to the time of concen-

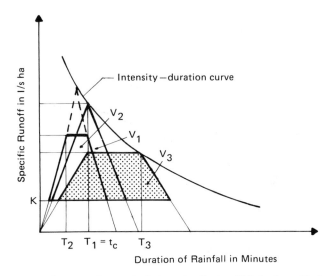

Figure 18.5   Storage volume calculation using Rational Method considering the time of concentration.

tration T_1, the storage volume is V_1 and is based on a triangular hydrograph that is symmetrical around T_1. When the duration is less than T_1, the volume is V_2 and is constructed by truncating the top of the triangle at a level where the recession limb intersects the time of concentration T_1. The triangle in this case is symmetrical around T_2. When the duration is greater than T_1, the volume is found under the trapezoid V_3. This trapezoid is constructed by drawing a horizontal line back from the value on I–D–F curve found at duration T_3 to T_1. The rising limb of the V_3 hydrograph is then a straight line connecting $T = O$ and the top of the trapezoid, and the falling limb is merely a mirror image of the rising limb.

A complete, step-by-step example of how the required volume is found, using metric units, is illustrated in Figure 18.6.

18.4 TIME–AREA METHOD

The time–area method was developed when computers where not as readily available as they are today. It is a method that has been incorporated into some of the common computer models such as British Road Research Model (BRRL) and Illinois Urban Drainage Area Simulator (ILLUDAS). It was developed to generate a more realistic runoff hydrograph than what is possible using the two Rational Formula procedures just described. In light of the available software on the market, The time–area method is not practical to use manually. The method is described here because it is incorporated into some

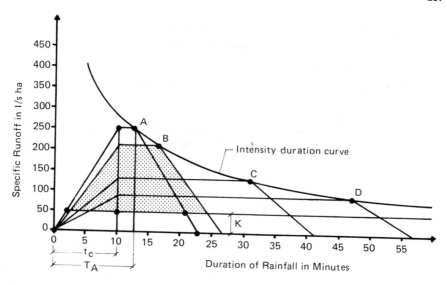

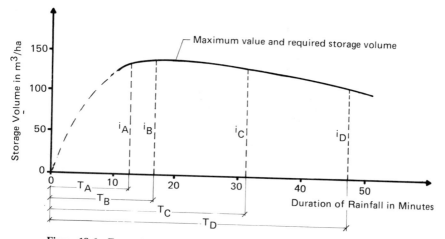

Figure 18.6 Example of the calculation of storage volume when considering time of concentration. (After Koral and Saatci, 1976.)

of the computer models on the market and also provides an example of one technique that can be used to construct a runoff hydrograph.

The time–area method received its name from the fact that the tributary watershed is subdivided into areas having similar flow times to the outlet. Using these subareas, a time–area curve is constructed using the following simplifying assumptions:

- each subarea remains constant throughout a storm;
- the time–area curve for each subarea is linear;
- the time of concentration is independent of the rainfall intensity; and
- the surface and pipe flow velocity remains constant throughout the storm.

The procedure for constructing a time–area curve for a large irregular watershed is illustrated in Figure 18.7. The drainage area is first divided into

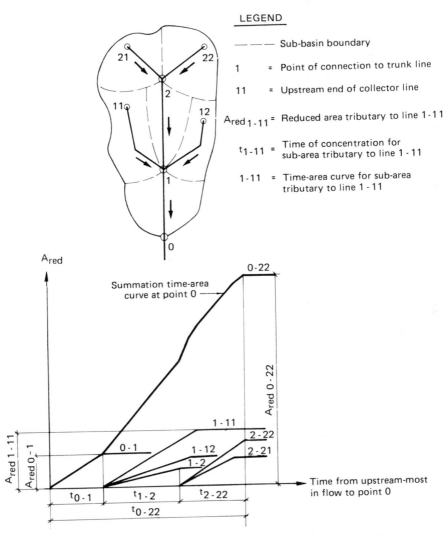

LEGEND

– – – –		Sub-basin boundary
1	=	Point of connection to trunk line
11	=	Upstream end of collector line
$A_{red\,1-11}$	=	Reduced area tributary to line 1-11
t_{1-11}	=	Time of concentration for sub-area tributary to line 1-11
1-11	=	Time-area curve for sub-area tributary to line 1-11

Figure 18.7 Constructing a time–area curve. (After Swedish Water and Sewage Works Association, 1976.)

a number of subwatersheds. For each subwatersheds, a time–area curve is plotted with its starting point lagged by the time it takes water to flow from the bottom of the subbasin to the bottom of the watershed. All of the subbasin time–area curves are then added to obtain the cumulative time–area curve for the entire watershed.

The flow rate at the bottom of the watershed for any storm duration can be found by multiplying the tributary area and the rainfall intensity at that duration by the runoff coefficient. You may have noticed that this technique sounds similar to the Rational Formula. In fact it is, since it uses block rainfall.

The design rain is assumed to begin at time zero of the time–area curve and has a duration equal to the time of concentration for the watershed. Obviously, after the rain stops, the effective tributary area of the watershed begins to shrink as the surface water is drained off. This method assumes that the runoff velocity after the rain stops remains the same as when the rain is falling. Thus, the time–area curve for the diminishing tributary area phase of the storm is the same as for the increasing tributary area phase of the storm. A runoff hydrograph for any block storm with a duration greater than the time of concentration can be found using the procedure illustrated in Figure 18.8:

1. Plot a parallel time–area curve that is shifted to the right a distance equal to the storm duration.

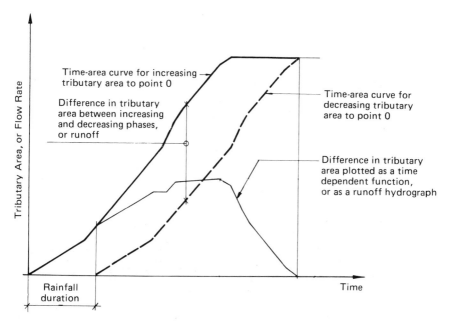

Figure 18.8 Constructing a time–area curve hydrograph. (After Swedish Water and Sewage Works Association, 1976.)

2. Subtract the ordinates of the shifted time–area curve (i.e., decreasing area phase) from the ordinates of the increasing time–area phase curve.
3. Multiply the difference in ordinates at each duration by the average rain intensity of the block design storm duration.
4. Multiply the values obtained in step 3 by the runoff coefficient for the watershed.

When the time–area method is used to size a detention facility, it is necessary to develop several runoff hydrographs for different storm durations. All of them are then analyzed to determine which will give the greatest storage volume. See Figure 18.9 for a graphic illustration of this process. The volume of storage is determined by measuring (i.e., numerically integrating) the area between the inflow hydrograph and the design release rate. However, note that the height scale in Figure 18.9, when converted to a flow hydrograph, will vary for each storm duration when it is multiplied by the block rainfall intensity. Figure 18.9 is, in fact, a representation of composite time–area curves for various storm durations before they are multiplied by rainfall intensity and the runoff coefficient.

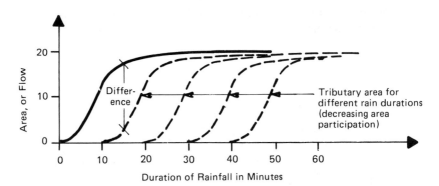

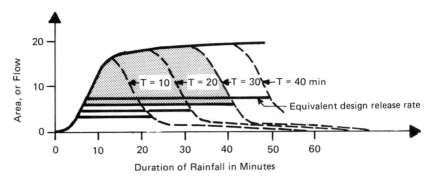

Figure 18.9 Constructing a time–area hydrograph for various storm durations. (After Swedish Association for Water and Sewer Works, 1976.)

18.5 RAIN POINT DIAGRAM METHOD

The methods described so far were based on the intensity–duration–frequency rainfall data. Although the intensity–duration data were originally intended for the calculation of peak runoff rates using Rational Formula, it is now commonly used to calculate storage volumes. This shift in the use of Rational Formula has been questioned by many and, as a result, the design of detention facilities using it has not always been accepted as accurate.

A method for estimating runoff volumes based on other types of rainfall data was suggested by van den Herik (1976). Van den Herik chose to base the calculations solely on the amount of total rainfall depth and the duration of recorded rainstorms. The method described next includes modifications suggested by Pecher (1978, 1980) which tailor it for the sizing of detention facilities.

18.5.1 Rain Point Diagram

This procedure requires the preparation of a rain point diagram by plotting total rainfall volume, the dependent variable, against the rainstorm duration, the independent variable for each recorded rainstorm (see Figure 18.10). To do this, it is first necessary to differentiate between individual rainstorm events in the data base.

Originally, van den Herik (1976) and Pecher (1978, 1980) suggested that individual storms be defined as separate events when the end of one storm is separated from the beginning of the next by at least one hour. Subsequent to this, Urbonas and Stahre, based on their own experience, developed modifica-

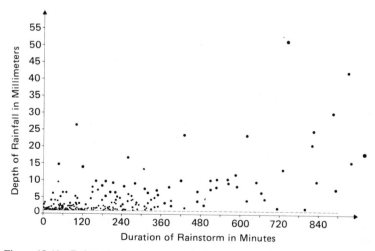

Figure 18.10   Rain point diagram for Stockholm, Sweden for 1978–79 period and 60 minutes between storms.

tions to these recommendations which will be discussed later in this chapter. Note that the rain point method ignores the temporal distribution of the rainfall intensity within a storm.

18.5.2. Estimating Rainfall Abstraction

When using the rain point diagram method for estimating storage volumes, it is also necessary to estimate rainfall abstractions that are likely to occur during all storms. In its simplest form, this can be done considering only the following:

- Runoff from unpaved surfaces can be assumed to be virtually zero. This assumption is reasonable for most rainstorms. It may not be accurate for storms that have large rainfall volumes or very high intensities.
- Initial losses are due to the wetting of pavement and entrapment in depression storage. These can range, according to Pecher (1969), from 0.02 inches for steep areas to 0.06 inches for flat areas. Viessman et al. (1977) reported depression storage on pavement to range from 0.02 inches for very steep areas to 0.135 inches for very flat areas.
- Evaporation losses. These can vary significantly between regions due to climatic conditions. For Holland, Pecher (1969) estimated evaporation losses from impervious surfaces at 1.0 to 1.6 liters per hectare-second (0.34 to 0.55 inches per day). Similar rates can be expected for many of the northeast coastal regions of the United States. Higher rates can be expected in southern regions and in the semi-arid regions of the western and southwestern United States.

18.5.3 Developing Abstractions and Discharge Envelope

Combining all of the rainfall abstractions and the storage facility discharge rates results in a diagram as shown in Figure 18.11. This diagram contains envelopes of volumes that should not require storage. Thus, by superimposing this diagram on top of the rain point diagram, it is possible to determine the numbers of rainstorms that will be stored in a storage basin of a known volume. All the points above the storage envelope represent storms that will exceed the facility's capacity and cause it to be overtopped.

18.5.4 Determining Storage Volume

As noted, the adequacy of any storage volume can be evaluated by superimposing the rainfall–abstractions–discharge envelope diagram onto the rain point diagram. The process of sizing the required storage volume can be

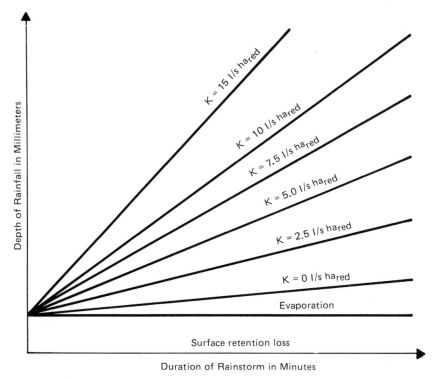

Figure 18.11   Rainfall abstractions and discharge envelope. (After Pecher, 1978.)

expedited by plotting the rainfall–abstractions–discharge envelope on a transparent overlay. This can be laid over the rain point diagram so that it can be parallel-shifted and various storage and/or release scenarios can be tested.

Shifting the overlay to the left a distance equal to the watershed's time of concentration (i.e., t_{run}) accounts for the watershed's time of concentration. Next, the overlay is parallel-shifted upward a distance corresponding to the storage volume of the detention facility (i.e., V) plus the surface retention loss.

An example of this procedure is illustrated in Figure 18.12. Using this procedure, various combinations of storage V and storage release discharge rates K can be quickly evaluated. All the points above the selected K line represent the storms that exceed the storage volume and result in an overflow condition. This analysis provides an indication of how runoff obtained using a long-term record of rainstorms interacts with the proposed design. Estimates of how frequently a given detention facility can be expected to be overloaded provides the designer information on how well local criteria or project goals are going to be met.

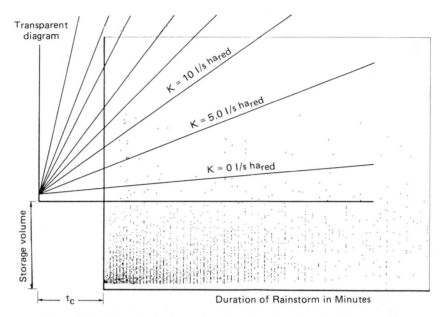

Figure 18.12   Example of volume determination using rain point diagram. (After Pecher, 1978.)

18.5.5 Summary and Suggested Update of Method

The foregoing method appears to have many advantages over traditional, Rational Formula-based methods for sizing of storage volumes. The main advantage is that actual rainfall data is used in the region instead of intensity–duration information, the latter being only a representation of rainfall and not actual storms as they may have occurred. However, the rain point diagram method has certain deficiencies and limitations as well.

For the rain point diagram to have a degree of reliability, it needs a continuous rainfall record of at least 15 to 20 years. This period is probably adequate only when designing storage for smaller than 10-year runoff events. Design for control of a 100-year runoff event will require in excess of 100 years of continuous data, while a 10-year event may be adequately estimated using 20 or more years of data. As a result, this method should be most reliable for the design of detention facilities for the control of frequently occurring rainstorms. These smaller storm events are of prime interest in the design of water quality treatment facilities.

The other major shortfall of this method is the arbitrary nature of separating rainfall data into discrete storm events. It does not take into account the effect of several successive rainstorms. As a result, the tendency will be to underestimate the storage volume for detention basins with low release rates, since the volume is not likely to be emptied between successive events. To compensate for this potential problem, the authors recommend the following when separating storms for the rain point diagram:

1. Determine the desired release rate from the storage prior to the rainfall data analysis.
2. Using the desired release rate, determine the time it will take for a full basin to be completely drained (T_D).
3. Determine the time of concentration for the watershed (T_c).
4. Separate storms using no less than 1-hour, $\frac{1}{2} T_D$, or T_c, whichever is the largest, as the separation interval between individual storms.

The preceding accounts for the rate of runoff from the watershed in relation to rate the water is released from the storage pond. For most small urban watersheds, the time of concentration is very small (i.e., approximately 10 minutes) and will not govern the storm separation period.

This procedure is only a partial remedy and a simplification for a very complex stochastic process. It should not be viewed as rectifying all shortcomings; however, it should reduce the probability of an undersized detention facility. The rain point diagram method with the preceding modification should result in reasonable storage volume estimates without performing continuous simulation modeling.

18.6 CUMULATIVE CURVE METHOD

18.6.1 Introduction

In some situations, the designer's goal is to equalize the daily variations in dry weather flow in a sewer system. These may occur because of the natural variations in dry weather flow or may be caused by short-term wastewater loads from an industry. Flow equalization volume in such cases is determined using the cumulative mass curve method. Cumulative mass curves are based on inflow measurements.

When developing cumulative mass curves, be aware that flow can vary within a given day, from day to day, from week to week, from month to month, and even over much longer periods of time. As a result, the designer has to decide over what period of time flow equalization has to be provided. This decision is typically based on flow observations over an extended period of time. As a minimum, one year's data has to be available before a reasonable flow equalization volume can be estimated. Even then, it is advisable to provide a safety factor in the final design.

An example of daily inflow variations is illustrated in Figure 18.13. In addition to flow, this figure also shows the measured variations in biochemical

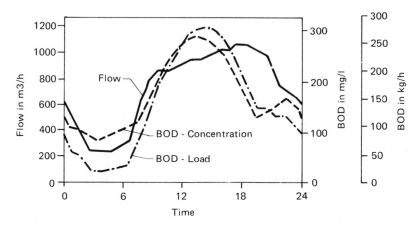

Figure 18.13 Example of daily flow variation at a wastewater treatment plant. (After EPA, 1974.)

oxygen demand (BOD). We describe the steps needed to estimate an appropriate flow equalization volume necessary to balance the flows over a one-day period.

18.6.2 Construction of a Cumulative Curve

The inflow hydrograph for the desired equalization period is transformed into a cumulative curve by

1. subdividing the entire flow equalization time period into uniform time increments, such as one-hour increments;
2. calculating the volume of inflow for each time step; and
3. accumulating all the incremental volumes over time.

The resultant cumulative volume table is then plotted against time, resulting in a cumulative mass curve similar to the one illustrated in Figure 18.14. This curve is based on the flow data shown in Figure 18.13.

The slope of the straight line connecting the beginning and end points of the cumulative mass curve represents the average flow rate over the period. The portions of the curve with a greater slope than the connecting line represent periods when the flow rate is greater than average. Similarly, portions of the curve with lesser slope than the connecting line represent periods of less than average inflow rate.

18.6.3 Determination of Storage Volume

The flow equalization volume is found using the cumulative mass curve as follows:

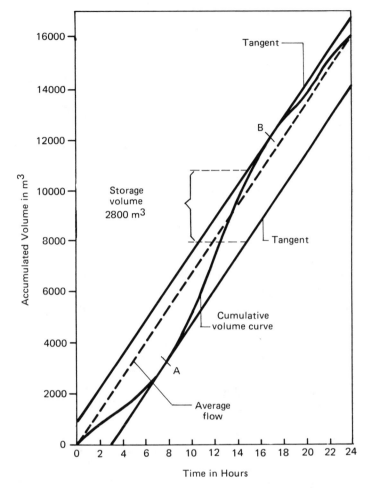

Figure 18.14   Storage volume determination using cumulative curve. (After EPA, 1974.)

1. Draw tangents to the curve parallel to the line representing the average flow (see Figure 18.14).
2. At the lower point of tangency *A*, the average flow rate begins to be exceeded and storage begins to fill.
3. The inflow is greater than the outflow until the upper point of tangency is reached at *B*.
4. The vertical distance between the two tangent lines is the required minimum design storage volume.
5. In practice, the design volume is set somewhat larger than determined in step 4 to provide a safety margin against unforeseen flow variations.

18.7 DETAILED CALCULATION METHODS

Detailed calculation methods encompass a family of computer rainfall runoff simulation models that have become readily available to any engineer in recent years. These models can simulate runoff, with or without water quality, on the basis of watershed characteristics, conveyance system characteristics, and recorded rainfall. If recorded rainfall is not readily available, the models can also be used with design storms. Although these models perform a large number of hydraulics calculations each second, their reliability and accuracy depends entirely on calibration against recorded simultaneous rainfall and runoff data.

There is a large number of models on the market, most of which will run on a personal computer. Any of them can be used to design and or analyze stormwater detention. However, we limit our discussion to the following three:

- Illinois Urban Drainage Area Simulator (ILLUDAS),
- EPA Storm Water Management Model (SWMM), and
- Urban Drainage Storm Water Model Version 2 (UDSWM2-PC).

These models were selected for description because all three are in public domain, are well-documented, and are supported by governmental agencies. Also, all three are very inexpensive to acquire.

These models are designed to simulate the entire runoff process, on the surface and inside storm sewers. Stormwater detention simulation routines are only a small part of each of the computer models. The performance of the three is described in greater detail in Chapter 19.

The main advantage of the detailed calculation methods is that the design no longer has to be based on block rainfall or the I–D–F curves. Any conceivable rainstorm distribution can be used, or the runoff from an entire chronologic series of recorded rainstorms can be studied. Pecher (1980), after studying several locations in West Germany, reported significant differences in detention sizing results when using continuous simulation instead of I–D–F-based block rain design storms.

The results of the studies by Pecher in West Germany show that

- traditional design methods underestimate storage volumes for facilities with small specific outlet capacities; and
- these same methods overestimate the storage volume for facilities with large specific outlet capacities.

Underestimates in detention volumes for facilities with low release rates K are attributed to successive storms arriving before the storage basin is completely drained. On the other hand, the overestimating of volume for facilities

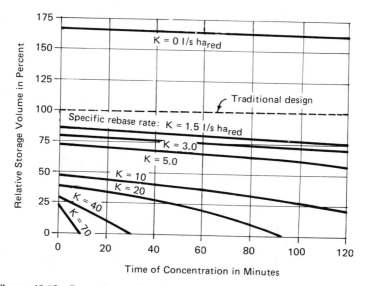

Figure 18.15 Comparison of storage volume calculated using continuous simulation and "traditional" methods. (After Pecher, 1980.)

that empty rapidly is attributed to the fact that design storms rarely represent storms that actually occur in nature. They are biased toward overestimating average intensities because of the fact that they are a composite of the most severe portions of many storms. It was also observed that the overestimating of storage volumes increased with increasing tributary area and specific release rate from the storage basin.

Some of Pecher's (1980) findings are presented in graphical form in Figure 18.15. In this figure, the storage volume obtained using traditional methods based on the Rational Formula was set equal to 100%. As can be seen, the variation in storage volume was found to be very significant. This is a compelling argument for similar investigations in the United States. Although it may be premature to draw general conclusions, Pecher's studies indicate that we may be in fact building thousands of detention basins that may either be wasteful or inadequate in size.

REFERENCES

ANNEN, G., AND LONDONG, D., "Comparative Contributions to Dimension-Determining Procedures for Detention Basins," Technisch-Wissenchaftlicht Mitteilungen der Emschergenossenschaft und des Lippeverbandes, No. 3, 1960. (In German)

APWA, *Practices in Detention in Urban Stormwater Runoff,* Special Report No. 43, American Public Works Association, Chicago, Ill., 1974.

APWA, *Urban Stormwater Management,* Special Report No. 49, American Public Works Association, Chicago, Ill., 1981.

DRCOG, *Urban Storm Drainage Criteria Manual,* Denver Regional Council of Governments (Currently being published by the Urban Drainage and Flood Control District), Denver, Colo., 1961.

EPA, *Flow Equalization,* Technology Transfer Seminar Proceedings, May, 1974.

FAA, *Airport Drainage,* Federal Aviation Agency, Washington, D.C., 1961, revised 1966.

HERIK, A. G. VON DEN, "Water Pollution by Storm Overflow From Mixed Sewer Systems," Berichte der ATV, No. 28, Bonn, 1976. (In German)

KAO, T. Y., "Hydraulic Design of Stormwater Detention Basin," Mini Course No. 3, National Symposium on Urban Hydrology and Sediment Control, University of Kentucky, July, 1973.

KELLY, H. "Designing Detention Basins for Small Land Developments," *Water and Sewer Works,* October, 1977.

KOOT, A. C. J., "Storage and Runoff Capacity of Mixed Water Sewers and Their Effects on Drainage and Clarification Plants," *Netherlands Calculation Methods,* Berichte der ATV, No. 23, Bonn, 1969. (In German)

KORAL, J., AND SAATCI, A.C., *Rain Overflow and Rain Detention Basins,* 2nd edition, Zurich, 1976. (In German)

KROPF, A., AND GEISER, A., "Detention Basins and Storm Water Clarification Installations," Schweizerische Bauzeitung No. 18, 1957. (In German)

LAUTRICH, R., "Graphic Determination of Storage Volume for Flood Containers, Seepage and Rain Equalization Basins and the Limiting Values of the Retardation," Wasser und Boden No. 8, 1956. (In German)

MACINNES, C., MIDDLETON, A., AND ADAMOWSKI, K., "Stochastic Design of Flow Equalization Basins," *Journal of the Environmental Division,* ASCE, December, 1978.

MALPRICHT, E., "Planning and Construction of Storm Water Detention Basins," Berichte der ATV, No. 15, Landesgruppentagungen, 1962. (In Swedish)

METCALF & EDDY, INC., *Wastewater Engineering, Treatment, Disposal, Reuse,* 2nd edition, McGraw-Hill, 1979.

ORDON C., "Volume of Storm Water Retention Basins," *Journal of the Environmental Division,* ASCE, October, 1974.

OUANO, E. A., "Developing a Methodology for Design of Equalization Basins," *Water and Sewer Works,* November, 1977.

PECHER, R., "The Runoff Coefficient and its Dependence on Rain Duration," Berichte aus dem Institut fur Wasserwirtschaft und Gesundheitsingenieurwesen, No. 2, TU Munich, 1969. (In German)

PECHER, R., "Design of Storm Water Retention Basins," NORDFORSK Report, Seminar on Detention Basin, Marsta 7–8 November, 1978. (In Swedish)

PECHER, R., "Dimensions of Storm Water Detention Basins According to Modern Rain Evaluation," Koncept 1979. (In German)

PECHER, R., "Extensive Storm Evaluation for Designing Storm Water Detention Basins," Abwassertechnik No. 1, 1980. (In German)

STAHRE, P., "Superficial Calculations of Storage Volumes," Royal Institute of Technology, Stockholm, Sweden, 1979. (In Swedish)

SWEDISH WATER AND SEWER WORKS ASSOCIATION, *Calculation of Detention Basins,* Publication P31, 1976. (In Swedish)

SWEDISH WATER AND SEWER WORKS ASSOCIATION, *Calculation of Sewer Networks,* Publication P28, 1976. (In Swedish)

VIESSMAN, W., JR., KNAPP, J., AND LEWIS, G. L., *Introduction to Hydrology,* 2nd edition, Harper & Row, p. 69, 1977.

YRJANAINEN, G., AND WARREN, A., "A Simple Method of Retention Basin Design," *Water and Sewer Works,* December, 1973.

19

Overview of Several Computer Models

19.1 GENERAL

Since the advent of the personal computer (i.e., PC), there has been an explosion of computer model use in water resources engineering. Today, there is a wide variety of stormwater system design models, most of them at very inexpensive prices. Because of this variety in available models, we limit our discussion to three models which we have personally used.

All three models are in public domain and are associated with a government agency. Unlike most of the software offered by private vendors, the source code for these three models can be obtained from each of the sponsoring agencies, examined, and modified by the user. We do not discourage the use of proprietary models, since many of them are excellent. We are not in the position however, to explain how proprietary models perform their calculations, or if they in fact have been adequately tested under a wide range of design conditions.

Computer models provide the ability to mathematically describe the performance of the entire stormwater system in much greater detail than was possible using hand calculations. This should not be misinterpreted that answers obtained using computer models are more accurate. In fact, it is too easy to have the model do the work without questioning the validity of the results. Computer models are only as good as the user and his or her experience. They cannot think for the user.

We suggest that you become totally familiar with the models you wish to use. Once you start using any model, always inspect its output very carefully to insure that the model, in fact, is giving representative results. As a rule, always calibrate your model either against field data or, lacking data, against a regionally accepted calculation technique. At least in the latter case, the results will be consistent for the location in which you are working.

The fact that computer models are now common should not exclude from your consideration other methods of calculation. Very often, it is sufficient to use simple hand calculations to estimate flows or to size a detention facility. This is particularly true for small urban watersheds. Unless sufficient rainfall and runoff data are available, single detention facilities can often be sized accurately using simple hand calculations. Remember, design storms do not give you an exact representation of what happens in nature, and processing them through a sophisticated model will not improve the accuracy of the results. Models do, however, give you an edge in comparing the effects of system performance as various proposed system components are tested for relative performance.

19.2 ILLUDAS

ILLUDAS stands for Illinois Urban Drainage Area Simulator. It has an option for the sizing of storm sewers given the basin runoff characteristics, design rainstorm, and the layout of the sewer network. If the sewer sizes are already known, such as in an existing system, the program will calculate the flows within the entire sewer network.

The model was first developed during the 1960s at the Road Research Laboratory in England and was referred to as the RRL method. It was further developed and enhanced by the Illinois State Water Survey and, since it was in public domain, it was made available to anyone by the state of Illinois upon request. In recent years, this model was converted to a PC version by two individuals working for the Illinois Water Survey and is being distributed as a proprietary model outside of Illinois.

ILLUDAS includes routines for estimating detention storage volumes. One of these routines is a simplification of the flood routing process occurring at a stormwater detention facility. This simplified routing option in ILLUDAS should only be considered for preliminary presizing of volumes before serious and more detailed studies are initiated. We refer to this preliminary routing procedure whenever ILLUDAS is discussed. For more information on the model and its capabilities, contact the Illinois State Water Survey.

19.2.1 Calculation Principles

ILLUDAS calculates flows within any sewer network using the normal depth technique. Like most other computer models, it does not account for backwater effects occurring in the sewer system. As a result, the flow capacity in a sewer pipe is reached when it is flowing full. If the flow entering the sewer exceeds the pipe-full capacity, the excess is accounted for by "storing" it at the upstream end of the sewer until the pipe flow decreases and capacity becomes available to "empty" the "storage." Note that we put quotes around "storage" and "empty" to emphasize that these processes are, in fact, only mathematical and are used to account for the mass balance in the system. Storage and emptying may not be actually taking place physically in the drainage system. The excess flow may instead be flowing overland in streets, yards, parking lots, etc.

The principle of storage and emptying in ILLUDAS is illustrated as a hydrograph in Figure 19.1. ILLUDAS reports the accumulated excess water in the output printouts. As a result, the user may compare the computed results against system constraints or desired results.

19.2.2 Detention Calculations

When designing a new stormwater sewer system using ILLUDAS, it is possible to incorporate detention facilities at any location within the sewer network. New detention facilities can be sized in two ways:

Available Storage Volume is Specified.   ILLUDAS will size a circular outlet pipe from the facility. The longitudinal slope and Manning's roughness for the outlet pipe must be given. The pipe is sized, using pipe-full flow to achieve exactly the stated available storage volume using the design runoff hydrograph.

Maximum Allowable Outflow Rate is Specified.   The model calculates the detention volume using the specified maximum outflow rate. In this op-

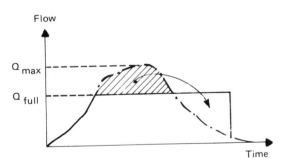

Figure 19.1   Illustration of storage calculation in ILLUDAS. (After Terstriep, 1974.)

tion, the maximum release rate can be specified either as a numeric value or the characteristic of the outlet pipe, namely pipe geometry, slope, and Manning's roughness coefficient.

Existing sewer networks may be analyzed either to determine the adequacy of existing detention facilities or to learn how much additional storage volume is needed. If the discharge exceeds the capacity of the outlet pipe, ILLUDAS calculates the required storage volume, which can be compared to the available volume. As an alternative, the available storage volume can be used as the limiting factor and the required outlet pipe size can be estimated. All flows are determined using normal depth calculations and do not account for surcharge of the outlet. As a result, the general tendency for this type of an algorithm is to underestimate the required volume or the size of the required outlet pipe. In effect, this algorithm simulates an off-line storage basin.

19.2.3 Summary

Detention design using ILLUDAS is performed using certain simplifying assumptions. Of these, the most significant is that the outflow from the detention facility is held constant during the entire detention process, namely, during filling and emptying. This simplification limits the use of ILLUDAS to preliminary systems planning. Figure 19.2 illustrates a hypothetical installation that approximates the detention model used by ILLUDAS.

19.3 SWMM

SWMM stands for Storm Water Management Model. It was developed and is being distributed under the sponsorship of the U.S. Environmental Protection Agency. It appears to be the most comprehensive urban stormwater network analysis model in practical application today. It was originally introduced in

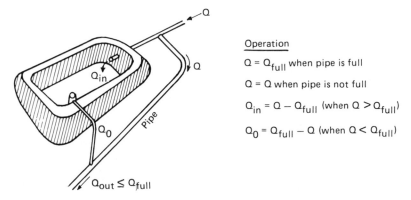

Operation

$Q = Q_{full}$ when pipe is full

$Q = Q$ when pipe is not full

$Q_{in} = Q - Q_{full}$ (when $Q > Q_{full}$)

$Q_0 = Q_{full} - Q$ (when $Q < Q_{full}$)

$Q_{out} \leq Q_{full}$

Figure 19.2 Example of a hypothetical detention as modeled by ILLUDAS. (After Terstriep, 1974.)

1971 and has been updated several times since. Version III was released in 1981, which for the first time provided a continuous simulation option for this model. As a result, Version III and IV and subsequent updates can be run as single event models or as continuous simulation models. Practical considerations of time or rainfall data availability may require the use of a longer integration time increment when using the continuous simulation option. The model is available through the University of Florida in Gainesville, where it is maintained and supported under a contract with EPA, or from the EPA office in Athens, Georgia.

Since its original release, SWMM was modified for specific applications, or for proprietary marketing by many different organizations. One version of it was modified for the province of Ottawa, Canada and was released as OTSWM. Another version was released as a proprietary PC compatible version by a professor at University of Hamilton in Canada. Currently, this version continues being marketed by the same group, but it is no longer associated with the University of Hamilton.

The Runoff Block of SWMM was also extensively modified by the Missouri Division of the Army Corps of Engineers. This version became known as the MRD Version of SWMM. The MRD Runoff Block has many of the flow routing options normally found in the Transport Block of the EPA version and also has some features not found in the EPA version of the Runoff Block. The MRD model offers a single package with many of the options frequently used in urban stormwater hydrology. In its current version, it is not recommended as a model to be used for estimating the runoff and transport of urban runoff pollutants. If storm runoff water quality needs to be modeled, the EPA version appears to be more appropriate. A modified version of the MRD SWMM is available from the Urban Drainage and Flood Control District (UDFCD) in Denver, Colorado as the UDSWM2-PC model.

The following discussion is limited to the EPA Version III SWMM program package. The UDSWM2-PC version is further described in section 19.4. Other versions are not described further here, and the reader may wish to contact the institutions we have mentioned or others offering their own versions for further information.

19.3.1 General Description of SWMM

The SWMM program consists of the following six blocks (see Figure 19.3)

- runoff,
- transport,
- extended transport (EXTRAN),
- storage/treatment,
- receiving water, and
- executive.

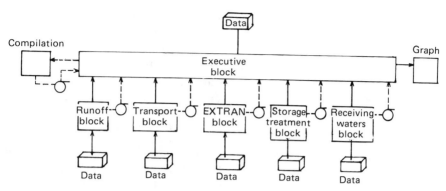

Figure 19.3   SWMM program blocks.

All of these blocks are not used simultaneously. Only the blocks best suited for a specific task are used at any given time. Output from one block can be used as input for another. This provides SWMM great flexibility and a staged approach toward modeling complex systems. Because of its extensive capabilities, the model obviously is very complex and is not one to be used casually. Effective use of SWMM is only possible after some training and after some experience has been gained by the user. For more details about this model, refer to the support manuals and publications for the Storm Water Management Model (1971, 1975–77, 1981, 1988), and subsequent updates.

Runoff Block.   The Runoff Block is used to estimate stormwater runoff from various subwatersheds, and its output can be used as input to the transport, EXTRAN, storage/treatment, or receiving blocks. Initial storm runoff calculations are based on sheet flow kinematic wave principle for the water that is not lost due to infiltration and surface retention. Any temporal and spatial distribution of rainfall can be used as input. Version III and subsequent updates permit continuous simulation of a chronologic series of rainfall and dry weather flow. Sheet flow, including the simulated pollutant load, is intercepted by trapezoidal gutters and circular pipes, which are then combined with flow and pollutants in other gutters and pipes. All flows and pollutants are eventually routed to specified discharge points. It is not necessary, however, to simulate pollutant runoff in order to use the runoff block.

Transport Block.   The Transport Block simulates the flow and pollutant transport in the major sewers of the system. Input data for the Transport Block consists of the output from the Runoff Block. This block can also simulate detention facilities at any point in the system. The calculations are based on the normal depth and continuity principle, which means they do not account for backwater effects or surcharge. If the inflow into any sewer segment exceeds its pipe full capacity, the excess is temporarily stored at the upstream end of the pipe segment. This algorithm is identical to the one we described for ILLUDAS and has a tendency to underestimate needed detention volumes.

Extended Transport Block (EXTRAN). By replacing the Transport Block by EXTRAN, it is possible to account for backwater effects in the flow conveyance system. The pressure gradient can go up to the ground surface at the nodes of the model. When the incoming flows surcharge the system so that it reaches the surface, the excess flows are not returned to the system. As a result, continuity is not maintained when a sewer system is surcharged excessively. EXTRAN also provides for the simulation of certain standard facilities such as overflows, pumping stations, detention facilities, etc.

Storage/Treatment Block. This block permits for simplified simulation of a single treatment plant in the system. The plant, however, has to be located at the downstream end of the sewer network. The treatment plant can include a single detention storage basin.

Receiving Water Block. The Receiving Water Block was designed to simulate the hydraulics and the fate of pollutants in the receiving bodies of water such as rivers, lakes, estuaries, etc. We do not discuss this block any further.

Executive Block. This block has the task of coordinating the information and transferring data between all of the other blocks in SWMM.

19.3.2 Detention Calculations in Transport Block

The Transport Block in SWMM III can be used to approximate in-line and off-line detention storage in the sewer system. At most, two storage basins can be simulated by this block (see Figure 19.4). If there are more than two basins, the system has to be broken up into smaller subsystems which can be

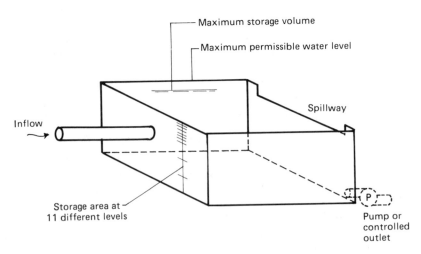

Figure 19.4 Detention facility as defined in Transport Block. (After Storm Water Mangement Model, 1971.)

simulated sequentially using the results of the upper network as input into the lower sewer network. The input data needed for storage calculations will include the following:

- Type of outlet structure. The choices provided by the model include bottom orifice outlet, constant rate pump, and a spillway.
- Depth–area relationship for up to 11 different water levels. This may be simplified in the case of a storage basin having the shape of an inverted circular truncated cone, in which case the user inputs only the bottom area and the slope of the walls.
- The maximum water level.
- The water level and the discharge rate at the start of the simulation.

The following three equations are used to describe the discharge through each type of outlet:

$$Q = A \cdot K_1 \cdot H^{1/2} \qquad \text{bottom orifice outlet} \qquad (19.1)$$

$$Q = L \cdot K_2 \cdot (H - h)^{3/2} \qquad \text{spillway} \qquad (19.2)$$

$$Q = K_3 \qquad \text{constant rate pump} \qquad (19.3)$$

in which Q = discharge rate,
H = depth of water above basin bottom,
A = area of the orifice outlet,
K_1 = constant dependent on orifice configuration,
L = length of spillway,
K_2 = constant dependent on spillway configuration,
h = height of spillway crest above basin bottom, and
K_3 = constant pump capacity.

When the pump option is used, it is also necessary to input the levels at which the pump is turned on and off.

If the water level in the storage basin during simulation rises above the maximum permissible level, the excess is not routed through the storage basin. Instead, it is accounted as excess volume in the printout of the simulation. This way, the modeler is aware of how much the basin may have been overloaded.

The pollutants in the system can also be routed through the storage basin. The program can estimate the removal of the settleable pollutants within the storage basin. This simulation can be performed, at the user's option, using plug flow or total mixed flow assumptions.

As a result, the program provides the modeler with a simulated hydrograph and a pollutograph after they are routed through the detention basin. Also, for each time step, the output provides the water depth and storage volume. The program does not provide a hydrograph of the water that may exceed the storage capacity of the facility and may spill as uncontrolled overflow.

19.3.3 Detention Calculations in Storage/Treatment Block

The SWMM program permits simulation of a treatment plant located at the downstream end of the system. Simulation of the following treatment plant components and processes is possible:

- gratings,
- swirl concentrator,
- sand trap,
- flotation,
- strainer,
- sedimentation,
- filtration,
- biological treatment, and
- chlorination.

The modeler excludes those treatment steps which are not applicable and provides the necessary basic parameters for the processes to be used. A storage facility can be located in-line or off-line to the sewer pipe entering the plant (see Figure 19.5). It is possible to use other connection schemes of detention and treatment plant than those shown in this figure. For example, when the storage is connected off-line, it is possible to route or pump the water from the storage basin to the plant.

Simulation of detention in this block is done using the same mathematic equations as used in the transport block described earlier. The only difference is that in the Storage/Treatment Block, the user has to specify the treatment efficiency for pollutant removal in the detention storage facility.

19.3.4 Detention Calculations in the EXTRAN Block

In the EXTRAN Block, the sewer network is represented by a series of links which are connected to each other at nodes. The modeler provides geometry, roughness, and invert elevations for each pipe. The user also has to provide the ground surface elevation at each node (i.e., manhole). Detention is simulated simply by providing the geometry of a pipe that best describes the storage vs. volume relationship of the installation. If the storage facility has an unusual shape, its characteristics can be approximated using any combination of pipes connected in parallel and series. The pipe sections that are supplied by the program are illustrated in Figure 19.6. The user may, however, describe additional pipes having any desired geometry.

It is possible to simplify the initial testing of a potential detention storage

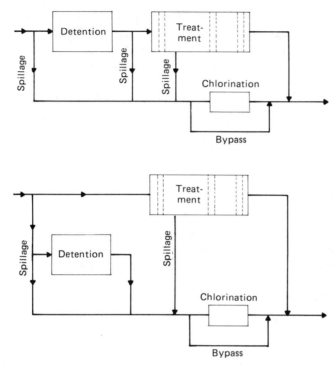

Figure 19.5 Treatment processes simulated by Storage/Treatment Block of SWMM.

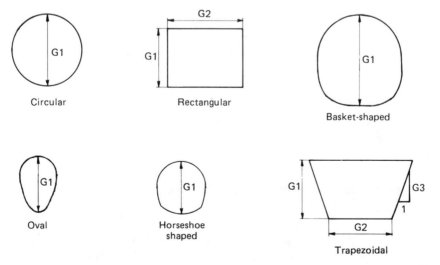

Figure 19.6 Standard pipe sections provided in EXTRAN Block. (After Storm Water Management Model, 1975–77.)

site without going into great geometric detail of the facility. This is done by defining node storage basins. All that is needed is to input the water surface area available at the node in question. EXTRAN assumes that the surface area remains constant as the water rises and falls and calculates the volume being stored at the node.

The outflow from a storage basin in EXTRAN is described by either giving the dimensions of the outlet pipe or one of the following flow regulating elements:

- overflows,
- outlet orifices,
- pumps, and
- high-water gates.

When these regulation elements are used to describe the discharge characteristics between two nodes, the user has to enter their hydraulic characteristics, e.g., discharge coefficients, spillway lengths, pumping rates, etc. An example of how a detention basin can be simulated using links, nodes, and flow regulating elements is illustrated in Figure 19.7.

The output from the EXTRAN block can provide for each time step the flow velocities in all the pipes and the water levels at all the nodes in the sewer network. At each of the detention sites, the inflow hydrograph, the outflow hydrograph, and the water levels are interdependent and are calculated simultaneously for each time step. This block is not simple to use, since all the component parts of the sewer system have to be described in detail and the calculations tend to become unstable if the element lengths are too short. It is a powerful tool for the analysis of an existing system and for the testing of proposed designs. It is not, however, the block that one would use for the general screening of many alternatives during planning.

19.3.5 Summary of SWMM Model Description

Detention calculations can be performed by the SWMM program in the following blocks:

- Transport Block,
- Extended Transport Block (EXTRAN), and
- Storage/Treatment Block.

The same mathematical equations are used in detention calculations in the Transport Block and the Storage/Treatment Block. The latter block only permits the simulation of detention at a treatment facility. Backwater effects are not considered in the Transport Block. If backwater effects are of signifi-

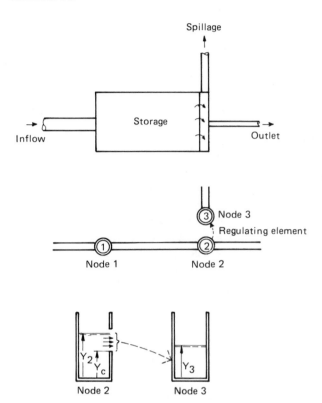

Figure 19.7   Example of how detention can be described using links, nodes, and flow regulators in EXTRAN Block. (After Storm Water Management Model, 1975–77.)

cant concern, the Transport Block can be replaced by EXTRAN, which accounts for water surface levels in the entire system.

Using SWMM, one can simulate most of the urban storm runoff and routing processes. It is a comprehensive and a powerful model and can be an extremely valuable tool in experienced hands. However, the model is complicated and imposes many requirements on the user. It is not a model of choice for casual investigation of what detention requirements may be needed at a single site. It is the model of choice for analyzing the performance of complete storm sewer systems, which may include detention facilities within such systems.

19.4 UDSWM2-PC

As mentioned earlier, the Runoff Block of SWMM was modified by the Missouri Division of the Army Corps of Engineers, which version was further modified to run on a PC for the Urban Drainage and Flood Control District

(UDFCD) in Denver, Colorado by the Boyle Engineering Company. In rewriting it for the UDFCD, the surface runoff calculations from tributary subbasins were decoupled from the gutter, pipe, detention, and other flow routing calculations. As a result, the user needs to generate the subbasin runoff hydrographs only once and then use them as input in subsequent runs. Various flow routing options can thus be studied at considerable savings in computer run time.

The user may also choose to generate storm runoff hydrographs using the UDFCD's Colorado Urban Hydrograph Procedure (CUHP) program. The output hydrographs from the CUHP program are then read by UDSWM2-PC, which routes these hydrographs through the conveyance system, detention facilities, diversions, etc.

This program has many of the routing options normally found in the Transport Block of SWMM. It also has some features not found in the Transport Block. Like the MRD Version of the Runoff Block, UDSWM-PC provides the following flow routing elements:

1. trapezoidal channels;
2. circular pipes;
3. direct flow links (i.e., no flow routing);
4. trapezoidal channels with an overflow channel;
5. circular pipes with an overflow channel;
6. detention facilities (based on Storage vs. Outflow rating table);
7. diversion facilities (based on Flow in Main Flow Element vs. Diverted Flow rating table); and
8. out-of-basin inflow hydrographs (based on Time vs. Flow table).

Also, like the MRD version of the Runoff Block, UDSWM-PC offers a single program block with many of the options frequently used in urban stormwater hydrology. In its current version, it has no capability to estimate the runoff and transport of urban runoff pollutants. If storm runoff water quality needs to be modeled, the EPA version of SWMM, despite its shortcomings in simulating pollutant loads, is the model of choice at this time. A feature was recently added to the new version of UDSWM that can automatically design the size of circular storm sewers.

19.4.1 Detention Calculations in UDSWM-PC

Detention calculations can be performed in two ways using UDSWM2-PC. The first option is an informal one and is similar to what we described for ILLUDAS. The user can obtain preliminary detention volume requirements by merely specifying a circular pipe of known flow capacity. The model will route the flows through the pipe until its pipe-full capacity is reached. Any excess flow is then held back in storage until the flows decrease and capacity in the pipe again becomes available to carry off the stored excess. The volume held back this way is reported along with the flow hydrograph and as the maximum volume stored in a summary table. Backwater effects and surcharge in the pipes are not considered in the calculations. As with ILLUDAS, the informal option produces estimated volumes that tend to be on the low side.

The second and formal detention option of UDSWM2-PC permits the user to define the outflow vs. storage characteristics for up to 25 detention facilities. The outflow vs. storage input data are used by the program only after the outlet pipe capacity is exceeded. In other words, the program will satisfy the normal depth capacity of the pipe element first before utilizing the outflow vs. storage tables provided by the user. This option permits an experienced user considerable flexibility in testing storage scenarios.

To simulate a surcharged outlet, the user enters the storage–outflow table and the characteristics for a very small pipe element that virtually has no flow capacity to satisfy. To approximate an off-line detention facility, the user specifies the pipe size equal to the bypass pipe and then enters the volume–outflow table for the flows that exceed its pipe full capacity. UDSWM2-PC is a single event model and will handle one storm event a time. Continuous modeling is not a currently available option.

The formal detention option calculates the storage in the basin using a Modified Puls flood routing procedure. The time increment used is the user-specified time increment of integration for all flow routing calculations in the model. The output consists of a printout that lists all the storage and discharge values throughout the run and the maximum discharge rate and volume stored throughout the storm. Full hydrograph values are printed only for the user-specified flow routing elements. A summary table of peak discharge rates and volumes stored, along with their respective times of occurrence, are printed for all routing elements within the model.

19.4.2 Summary

UDSWM2-PC is a modified version of the SWMM Runoff Block that will run on a PC. The modifications allow the user a variety of flow routing options, including detention facilities. Detention calculations can either be performed informally in a manner similar to how ILLUDAS handles them, or formally using the Modified Puls flood routing procedure. In the latter case, the user can specify up to 25 separate detention facilities anywhere in the flow routing network.

REFERENCES

STORMWATER MANAGEMENT MODEL, *SWMM User's Manual,* Volumes I–IV, EPA 11024 Doc 07-10/71, 1971.

STORMWATER MANAGEMENT MODEL, *SWMM User's Manual Version II,* EPA 670/Z-75-017, 1975. Supplemented 1976 and 1977.

STORMWATER MANAGEMENT MODEL, *SWMM User's Manual Version III,* EPA Project No. CR-805664, 1981.

TERSTRIEP, M. L., AND STALL, J. B., "The Illinois Urban Drainage Area Simulator, ILLUDAS," Illinois State Water Survey, Bulletin 58, 1974.

Urban Drainage Stormwater Management Model—PC Version (UDSWM2) Users Manual, Urban Drainage and Flood Control District, Denver, Colo., 1985.

20
Examples of Detention Basin Sizing

20.1 GENERAL

So far, we have described various methods to calculate detention basin volumes. These included the use of different design rainfall information and procedures to calculate storage volumes. It is up to the designer or planner to select the detention sizing methodology. In some cases, rough estimates of volumes and release rates are sufficient. In others, detailed system analysis is needed. We can arbitrarily subdivide the various sizing or analysis procedures into two general categories, namely:

- superficial sizing methods, and
- detailed calculation methods.

Superficial methods are of value during the initial planning phase of a project when rough estimates of volumes, land areas, and costs are needed. These methods, for the sake of jurisdictional consistency, are sometimes mandated by local authorities for final design as well. During the planning phase, it is not practical to use detailed procedures. At that time, the designer wants to assess many sites and facilities, and sizing calculations during this phase do not need to be very refined.

Detailed methods are reserved for the preliminary and final design phase. By the time the detailed methods are used, detention site locations have been identified and each of the facilities is designed or analyzed much more precisely. Detailed methods may also be used to evaluate the performance of entire stormwater collection, transport, and detention systems to improve the overall system performance.

20.2 EXAMPLES OF CALCULATIONS

Two examples are presented to illustrate the variability that is possible using various design procedures. The first example is based on experience in Sweden and is presented in metric units. The second example is based on experience in the Denver metropolitan area of the United States. This example is presented in English units. Both are simple single detention basin sizing examples which do not analyze the performance of an entire drainageway system.

20.2.1 Example from Sweden

In this example, a detention basin needs to be designed to limit the runoff from a 51.3 hectare (116 acre) residential watershed to 300 liters per second (10.6 cfs) during a one-year storm. The watershed has 20.8 hectares (47 acres) of impervious area; namely, the watershed has 40.5% impervious surface. The time of concentration was estimated to be 12 minutes.

As a first step, the detention volume was estimated using one of the superficial methods and block rain. To get an idea of how the results may vary due to the calculation method, calculations were first performed using a method that does not consider the time of concentration (see Figure 18.4). They were then repeated using a method that considers the time of concentration (see Figure 18.6). The results from both cases were then compared to the results obtained using the preliminary sizing option of ILLUDAS and block rain as input and are summarized in Table 20.1.

As can be seen, the calculated volumes are sensitive to the method of

TABLE 20.1 Comparison of Volumes Calculated Using Rational Formula Method, ILLUDAS, and Block Rain in m³

CALCULATION METHOD	RECURRENCE FREQUENCY (years)		
	0.5	1.0	2.0
Without time of concentration	988	1,380	1,903
With time of concentration	832	1,206	1,726
ILLUDAS	693	1,079	1,571

calculation. Both of the superficial methods appear to overestimate the storage volume when compared to ILLUDAS. However, we pointed out earlier that the calculating algorithm used in these runs of ILLUDAS has a tendency to underestimate detention volumes.

In Chapter 18, we mentioned that the block rain represents only a part of the total volume of a rainstorm. This is because it assumes that there is no rainfall before or after the central block rain. Sifalda and Arnell tried to compensate for this apparent shortfall through modifying block rain (see Figures 16.5 and 16.7). These same calculations were repeated using ILLUDAS and the Sifalda and Arnell design rainfall patterns, and the findings are compared in Table 20.2.

As can be seen from Table 20.2, the choice of design rainfall pattern can have as much, or even more, effect on the results than the choice of calculating method. The smallest volume resulted from the use of block rain, probably because it had the least total rainfall in the storm. The other two storm patterns were larger because of the so-called adjustments to make them more "representative" of real storms. At this time, there is no basis for judging which of the design storm patterns is most accurate in calculating storage volumes. Factors such as the detention period, the time of concentration of the watershed, and the duration of the design rainfall need to be considered before deciding which of these three storm patterns are most appropriate.

In all the calculations presented so far in Tables 20.1 and 20.2, the outflow rate from the storage basin was assumed to be constant at 300 liters per second (see Figure 20.1). In most detention facilities, the outflow rate will vary with the water depth. To illustrate how this variable outflow rate affects volume calculations, a more precise flow routing algorithm was used. This was done with the aid of KTH-UTMAG (Anderson and Stahre, 1981) computer model developed in Sweden. The same inflow hydrograph used in the ILLUDAS calculations and illustrated in Figure 20.1 was also used with the KTH model.

First, the storage volume obtained under the constant outflow rate assumption, namely 1,079 cubic meters shown in Figure 20.1, was fixed. It was also assumed that the storage basin was rectangular in shape with vertical walls. The outlet was assumed to be a vertical circular orifice. The inflow

TABLE 20.2   Comparison of Volumes Calculated Using Three Different Design Storms and ILLUDAS in m³

| | RECURRENCE FREQUENCY (Years) | | |
CALCULATION METHOD	0.5	1.0	2.0
ILLUDAS with Sifalda rain	1,092	1,627	2,308
ILLUDAS with Arnell rain	970	1,870	2,535
ILLUDAS with block rain	693	1,079	1,571

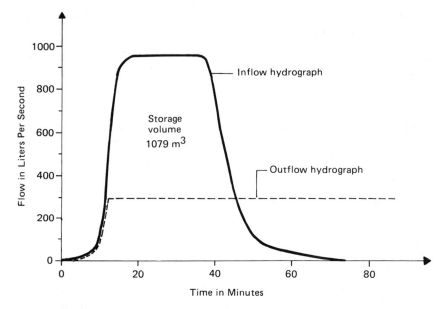

Figure 20.1   Results of ILLUDAS storage calculations under a constant 300 liter/second outflow rate.

hydrograph was then routed through the basin and an outflow hydrograph was calculated. The results are graphed in Figure 20.2, where we see that the actual peak outflow from a 1,079 cubic meter basin is 500 liters per second.

For the detention basin to maintain the 300 liter per second maximum outflow rate, the volume has to be larger. Next, the KTH model was used to find the detention basin volume needed to maintain the maximum outflow rate at 300 liters per second. The results of these calculations are shown in Figure 20.3. We see that the volume needs to be 25% larger than calculated previously to maintain the 300 liter per second release rate.

Obviously, the basin size is affected by how accurately the outflow vs. depth function is simulated. The lesson to be gained from these examples is that the use of a more sophisticated model ILLUDAS did not improve the accuracy of the calculations. In fact, neither one of the "superficial" methods estimated the final volume closer than ILLUDAS. This is not to say that ILLUDAS or similar models (i.e., Transport Block of SWMM, informal storage calculation procedure of UDSWM2–PC, or other similar procedures) should not be used. On the contrary, they are valuable tools during early planning as long as the user realizes that the estimates of storage volumes obtained using them will probably have to be adjusted upwards.

Another lesson from all this is to *know the algorithm that is used in the model you are using.* Blind faith in computer "black box" models has very little room in good engineering practice. Only through understanding of the various flow routing processes and the algorithms used to simulate them can

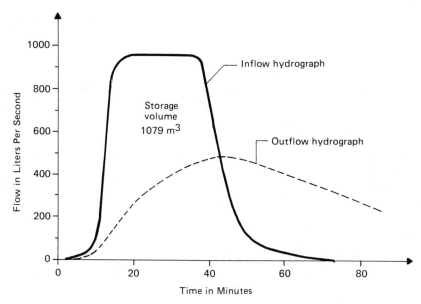

Figure 20.2   Results of detention basin calculations assuming a fixed volume shown in Figure 20.

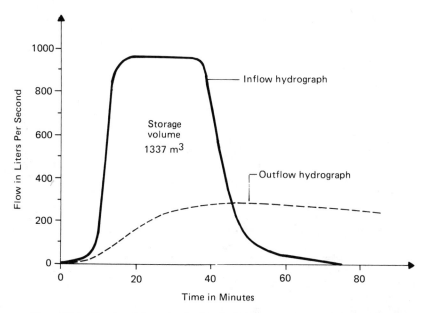

Figure 20.3   Results of detention basin calculations assuming a maximum release rate of 300 1/s.

the engineer be in position to judge which tools to use under various conditions.

20.2.2 Example from the United States

In this second example, a detention basin has to limit the runoff from a 100-year storm to 1 cubic feet per second per acre. The tributary watershed is a 100-acre residential development with 40% of its surface covered by pavement and rooftop. The watershed has a length to width ratio of two, a time of concentration of 32 minutes, and a runoff coefficient of 0.68 during a 100-year storm.

As we did for the first examples, the detention volume was first estimated using the Rational Formula storage calculation method without consideration for the time of concentration. The calculations for this method are tabulated in Table 20.3.

TABLE 20.3 Modified Rational Method for Detention Storage Calculations; Project Title: Example Problem for Denver Area

STORM DURATION (Min.)	RAINFALL INTENSITY (In./Hr.)	RUNOFF VOLUME (Cu. Ft.)	OUTFLOW VOLUME (Cu. Ft.)	STORAGE VOLUME (Cu. Ft.)	STORAGE VOLUME (Ac. Ft.)
(1)	(2)	$(3) = -60 \cdot (1)$ $\cdot (2) \cdot A \cdot C$	$(4) =$ $60 \cdot (1) \cdot Q$	$(5) =$ $(3)-(4)$	$(6) =$ $(5)/43,560$
5.0	8.82	181,403	30,000	151,403	3.48
10.0	7.03	289,383	60,000	229,383	5.27
15.0	5.90	364,244	90,000	274,244	6.30
20.0	5.11	420,819	120,000	300,819	6.91
25.0	4.53	465,999	150,000	315,999	7.25
30.0	4.08	503,483	180,000	323,483	7.43
35.0	3.72	535,458	210,000	325,458	7.47*
40.0	3.42	563,316	240,000	323,316	7.42
45.0	3.18	587,990	270,000	317,990	7.30
50.0	2.97	610,135	300,000	310,135	7.12
60.0	2.63	648,615	360,000	288,615	6.63
70.0	2.37	681,321	420,000	261,321	6.00
80.0	2.16	709,803	480,000	229,803	5.28
90.0	1.99	735,064	540,000	195,064	4.48
100.0	1.84	757,788	600,000	157,788	3.62
110.0	1.72	778,464	660,000	118,464	2.72
120.0	1.62	797,451	720,000	77,451	1.78
130.0	1.52	815,022	780,000	35,022	0.80
140.0	1.44	831,386	840,000	(8,614)	−0.20

* Required storage = 7.47 acre feet

Basin size:	100 acres
% Impervious:	40%
Runoff coef. (C):	0.68
Design frequency:	100–year
One hour rainfall:	2.6 inches
Design discharge:	100 cubic feet per second

Next, the storage volume was determined using the computer program HYDRO POND (Guo, 1987) a PC program developed at the University of Colorado at Denver to help design detention storage ponds. The inflow hydrograph for this example was generated using the Colorado Urban Hydrograph Procedure (CUHP) and its 100-year design storm. The use of the CUHP is of no particular significance here, and any other hydrograph generating method would have served as well for this example. It was used to keep the entire example consistent with the methods employed in the Denver region.

The results obtained using the modified rational method and HYDRO POND were then compared to the results obtained using the informal and formal flow routing process in UDSWM2–PC described in Chapter 19. The outlet hydrographs generated by UDSWM2–PC and HYDRO POND are shown in Figure 20.4.

Note that the informal process in UDSWM2–PC produces a similar outlet hydrograph shown earlier for ILLUDAS. Also note that the outflow hydrograph for the formal routing process in UDSWM2–PC and from HYDRO POND are almost identical. There appears to be approximately a 10-minute time shift between the two, but the shape, peak outflow, and the storage volume are identical. The results of all four calculations are summarized in Table 20.4.

Again, we see that the storage volume requirements vary with the method of calculation. In this example, the informal routing process, which approximates an off-line storage basin, gave the lowest volume. Specifically, it estimated a volume that was almost 20% lower than the two "exact" storage routing calculation methods. The superficial Rational Formula-based method

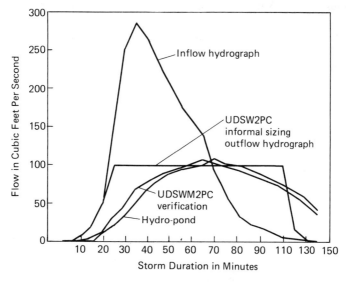

Figure 20.4 Comparison of hydrograph routing using UDSWM2–PC and HYDRO POND for an example in Denver.

TABLE 20.4   Volumes Calculated Using Rational Formula Method, UDSWM2–PC, HYDRO-POND, and CUHP Design Storm

	100-YEARS RECURRENCE FREQUENCY	
Calculation Method	Volume in Acre Feet	Peak Flow in Cubic Feet/Sec
Modified rational method	7.47	100
HYDRO-POND	8.21	106
UDSWM2–PC formal routing process	8.21	106
UDSWM2–PC informal storage sizing	6.7	100

estimated a volume that was within 10% of the volume found using the "exact" calculating methodology.

Based on the examples from Sweden and Denver, it is tempting to conclude that the so-called "superficial" detention sizing methods may in fact be superior to the Transport Block of SWMM and the informal options of UDSWM2–PC and ILLUDAS, or from other models using similar algorithms. Unfortunately, such a conclusion would be irresponsible. Before it can be reached, sufficient local studies need to be performed to determine which simplified procedures give the best preliminary results.

REFERENCES

ANDERSON, J., AND STAHRE, P., "KTH-UTMAG, Calculation Routine for Detention Facilities," KTH, Vattenvardsteknik, Rationella Avloppssystem, Medelande 17, 1981. (In Swedish)

GUO, C. Y., *User's Manual for HYDRO POND, A Personal Computer Software Program for Reservoir Routing and Outlet Structure Design,* University of Colorado at Denver, January, 1987.

— 21 —

Stormwater Pollutants

21.1 INTRODUCTION

In Part 4, we discuss the emerging technology of using stormwater detention for the removal of pollutants found in separate urban stormwater runoff. Although many different constituents can be found in urban runoff, to avoid being overwhelmed it helps to focus primarily on certain pollutants that can be used as representative indicators of others. Because of this, the EPA (1983) adopted for their Nationwide Urban Runoff Program the following constituents as " . . . standard pollutants characterizing urban runoff":

TSS	Total suspended solids
BOD	Biochemical oxygen demand
COD	Chemical oxygen demand
TP	Total phosphorus (as P)
SP	Soluble phosphorus (as P)
TKN	Total Kjeldahl nitrogen (as N)
$NO_{2\&3}$	Nitrite & nitrate (as N)
Cu	Total copper
Pb	Total lead
Zn	Total zinc

A number of other constituents were evaluated by the EPA before selecting the foregoing list. The EPA explains this selection as follows:

> The list includes pollutants of general interest which are usually examined in both point and nonpoint source studies and includes representatives of important categories of pollutants—namely *solids, oxygen consuming constituents, nutrients,* and *heavy metals.* [emphasis added]

21.1.1 National Urban Runoff Program

Although there were many studies of separate urban stormwater runoff quality, none was as extensive or included more sites than the U.S. Environmental protection Agency's National Urban Runoff Program (i.e., NURP). This program collected date during 1981 and 1982 which was analyzed and summarized in the final NURP report by EPA (1983). The final report, along with its technical appendices, provides much insight into urban stormwater runoff quality. It discusses the potential water quality standards violations in receiving waters and suggests the best management practices for reducing pollutant concentrations in stormwater runoff.

The final acceptable data base for this project came from 81 sites located in 22 different cities throughout the United States. It included more than 2,300 separate storm events. However, since all pollutants were not measured at all sites, the numbers of samples for individual pollutants were somewhat less than that. For the most part, the data consisted of the flow weighted average concentration, namely, the event mean concentration of each pollutant for each runoff event. In some cities, however, discrete samples were collected throughout the runoff event to characterize how the concentrations varied during any given event. These were then also flow weight composited to determine the Event Mean Concentration (i.e., EMC) for all storms.

21.1.2 Generalized NURP Findings

Before looking at the data summaries reported in the final NURP report, it helps to understand what the initial statistical analysis of the data revealed. Since the original NURP report was published, data analysis has continued and, as a result, there were follow-up activities by EPA and the United States Geological Survey to further refine the initial findings. It may be that some of the conclusions reported in the NURP report may yet be modified. However, it is unlikely that the broad observations reported in the final NURP report regarding data variability will be modified significantly. Briefly, EPA (1983) reported the following in its final NURP report:

- The EMCs at each test site were found to exhibit log-normal statistical distribution.
- The site median of the EMCs from all the test sites were found to also exhibit a log-normal statistical distribution. Figure 21.1 illustrates this for total copper (i.e. C_u).

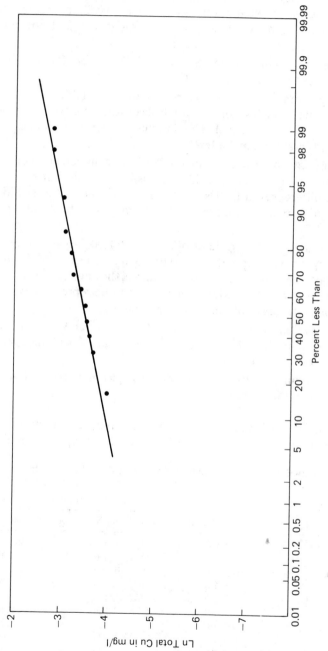

Figure 21.1 Log-normal distribution of total copper data collected under NURP. (After EPA, 1983.)

- For TSS, 90% of the individual storm EMCs were found to vary over a range of three to five times the site median EMC. Apparently, the variation in EMCs between storms can be significant, but not out of the ordinary for random data such as this.

- For constituents other than TSS, 90% of EMCs were found to be two to three times the site median EMC.

- Although some cities exhibited higher or lower EMC values than the national median EMC for one or more constituents, no clear geographical patterns were revealed. The national median EMC values of urban pollutants are listed in Table 21.1.

- Land use categories (i.e., residential, mixed, commercial, industrial and open/nonurban) do not provide a statistically significant basis for predicting differences in EMCs. Although not considered as significant by NURP, differences were observed and reported in the final report (see Table 21.2).

- Fecal coliform EMCs varied from 300 to 281,000 organisms per 100 ml during warm weather and from 20 to 3,300 organisms per 100 ml during cold weather. One cold weather sample, however, reported 330,000 organisms per 100 ml, which appears not to be representative and may have been the result of sample contamination.

- No corrolation was found between EMCs and runoff volumes, indicating that EMCs and runoff volumes are, for the most part, independent of each other.

- Runoff volume coefficient exhibited a logarithmic correlation to total basin imperviousness.

The NURP Final Report states that there are little, if any, statistically significant differences in constituent EMCs between geographic regions, be-

TABLE 21.1   Water Quality Characteristics of Urban Runoff

Constituent	EMC Coef. of Var. Event to Event	URBAN SITE MEDIAN EMC IN MG/L	
		Median Site	90th Percentile
TSS	1.0–2.0	100.000	300.000
BOD	0.5–1.0	9.000	15.000
COD	0.5–1.0	65.000	450.000
TP	0.5–1.0	0.330	0.700
SP	0.5–1.0	0.120	0.210
TKN	0.5–1.0	1.500	3.300
NO_{2+3}	0.5–1.0	0.680	1.750
Cu	0.5–1.0	0.034	0.093
Pb	0.5–1.0	0.140	0.350
Zn	0.5–1.0	0.160	0.500

After EPA, 1983

TABLE 21.2   Median EMC for All NURP Sites by Land Use Category

Constituent (mg/l)	RESIDENTIAL		MIXED		COMMERCIAL		OPEN/ NONURBAN	
	Median	CV	Median	CV	Median	CV	Median	CV
BOD	10.000	0.41	7.800	0.52	9.300	0.31	—	—
COD	73.000	0.55	65.000	0.58	57.000	0.39	40.000	0.78
TSS	101.000	0.96	67.000	1.10	69.000	0.85	70.000	2.90
Pb	0.144	0.75	0.114	1.40	0.104	0.68	0.030	1.50
Cu	0.033	0.99	0.027	1.30	0.029	0.81	—	—
Zn	0.135	0.84	0.154	0.78	0.226	1.10	0.195	0.66
TKN	1.900	0.73	1.290	0.50	1.180	0.43	0.965	1.00
NO_{2+3}	0.736	0.83	0.558	0.67	0.572	0.48	0.543	0.91
TP	0.383	0.69	0.263	0.75	0.201	0.67	0.121	1.70
SP	0.143	0.46	0.056	0.75	0.080	0.71	0.026	2.10

Note: CV stands for coefficient of variance.
After EPA, 1983

tween various cities, or between storm events at a given site. The results appear to be relatively uniform across the United States. However, the data reveals wide ranges in EMCs of all constituents. This finding tends to support the need for local data before making what may be far-reaching and expensive water quality management decisions.

21.2 SUSPENDED SOLIDS IN STORMWATER

We begin by discussing primarily total suspended solids (TSS) in stormwater, their characteristics, and the technology for their removal. It was the belief that the removal of TSS from stormwater would also be accompanied by a proportionate removal of other pollutants. This assumption was shown not to always be the case. Nevertheless, most pollutants appear to have a strong affinity to suspended solids, and the removal of TSS will very often remove many of the other pollutants found in urban stormwater.

Apparent exceptions to the assumption that the removal of TSS will also remove other pollutants are dissolved solids, nitrites and nitrates ($NO_{2\&3}$), and soluble phosphorus (SP). Dry detention basins, namely, basins that drain fully between storm events and have no permanent pool of water, have consistently exhibited very poor dissolved $NO_{2\&3}$ and SP removal efficiencies. Because the reductions in the $NO_{2\&3}$ and SP EMCs are difficult to achieve, facilities designed for their removal should provide significant reductions in the EMCs of other pollutants. Detention storage basins that have a permanent pool appear to provide the best efficiencies in reducing dissolved $NO_{2\&3}$ and SP concentrations.

21.2.1 Factors Affecting Settlement of TSS

As noted, stormwater quality appears to vary from one location to the next. Of significance are the type and quantities of pollutants found in stormwater and to what degree these pollutants may be associated with sediments. It is, therefore, a good idea to sample and analyze the stormwater that is to be treated in water quality enhancement basins. Among others, the following factors seem to be of significance in describing the settling characteristics of TSS and associated pollutants:

- pollutant load in the stormwater by type;
- the percentages of settleable pollutants;
- particle size distribution;
- distribution of the solids by their settling velocities;
- distribution of pollutants by settling velocities;
- particle volume distribution of the solids; and
- the density of the settleable pollutants.

21.2.2 First Flush vs. No First Flush

Pollutants enter stormwater in many ways, among which are the following:

- Pollutants are absorbed as the raindrops pass through the atmosphere.
- Pollutants are washed off the paved and unpaved surfaces by stormwater runoff.
- Pollutants that have accumulated since the last storm in sewers, ditches, and channels are picked up by the new stormwater. This source can be also aggravated by illegal wastewater connections to the stormwater conveyance system.

Some studies show that the pollutant concentrations are largest early in the runoff process. This is explained by speculating that the paved surfaces are most polluted before rain begins. As rainfall continues, the surface pollutant accumulation is depleted and pollutants are diluted by the larger flows in the transport system. Also, it is speculated that the so-called first flush also depends on the intensity and the duration of rainfall.

Studies at other sites have not found an identifiable first flush. At those sites, the pollutant concentration seems to have no relation to the duration of rainfall. This was the case in the Denver Urban Runoff Evaluation Project by DRCOG (1982). Some speculated that because Denver is in a semi-arid climate, the runoff never left first flush. This point of view is not supported by the data, since some of the storm events were in excess of 1.5 inches and had sufficient duration to clearly show if there was a strong first flush. Also, the EPA (1983) reported a nonexistent negative statistical correlation between EMCs and runoff volumes (i.e., $r^2 = 0.3$). If first flush was uniformly present, a much stronger negative correlation should have been found.

As a result of the conflicting findings, it is not appropriate to assume that by merely capturing the first flush most of the pollutants will be captured. In fact, lacking local data, it is safer to assume that there is no first flush. If local investigations do find a significant first flush, then it is necessary to define its volume and duration. It has been suggested that a strong first flush is present when 20% of the runoff contains 80% of the pollutants. This is not a truism, and the definition will vary between communities and investigators. Nevertheless, its quantification can significantly simplify the definition of the volume of runoff that has to be captured and treated.

21.2.3 Initial TSS Concentrations vs. TSS Removal

Stockholm's Water and Sewer Works (1978), namely their water works department, found that the rate of sedimentation is dependent on the initial concentration of TSS in stormwater. Figure 21.2 compares data of incoming TSS concentrations against the percent reduction in TSS in stormwater collected at a sedimentation tunnel near Stockholm. A clear trend supporting a correlation between incoming TSS concentrations and its removal can be seen in this figure. The same study also revealed that this installation could not reduce the TSS concentrations to lower than 10 to 20 milligrams per liter.

In contrast, recent laboratory settling studies in the United States by Randall et al. (1982) showed that even very small initial TSS concentrations in stormwater can be reduced further. Of course, these findings were for nondynamic conditions in laboratory settling tubes (see Table 21.3) and not for dynamic field conditions. The findings by Randall et al. support the Swedish findings that the amount of sediment removed through settling increases with increasing initial concentration of TSS.

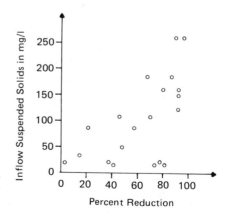

Figure 21.2   Percent reduction in TSS vs. the TSS concentrations in the inflow. (After Stockholm's Water and Sewer Works, 1978.)

TABLE 21.3   Time to Remove 60% of TSS at 4-foot Depth

Initial TSS, mg/l:	15	35	38	100	155	215	721
Settling time, hours:	38	24	8	5	1.0	1.5	0.5

After Randall et al., 1982

21.2.4 Easily Settling Pollutants

Currently, there is an interest to quantify the TSS in stormwater for the purpose of designing sedimentation facilities. Since many of the pollutants are attached to the solids, the removal of TSS will also remove many of the other pollutants. Unfortunately, as reported by Sartor and Boyd (1977), most of the pollutants are attached to the smaller size fractions of the TSS and are difficult to settle out. That is particularly a problem in the dynamic and turbulent environment found in a detention basin as it first fills with water and then is drained. Turbulence and wave action can maintain fine particles in suspension and even resuspend them from the bottom of the basin.

The settling ability of pollutants via the removal of TSS from stormwater involves the removal of different sized and density particles. The heavier particles will obviously settle out earlier than the lighter ones. As a result, there is a fraction of TSS that can be easily removed by simply holding the water for a short period of time in a forebay, which we call a *coarse pollutant chamber*. Such a chamber should be designed to detain the water for approximately five minutes during an average runoff event.

Particle weight by itself is not a total indicator of whether or not a particle will settle out rapidly. The proclivity of any individual particle to settle easily is also determined by the particle's effective density; namely, weight divided by total volume. As a result, some large particles may be difficult to remove through settling if their effective density is very low.

21.2.5 Particle Size Distribution of TSS

Laboratory particle size distribution and settling tests of urban pollutants were conducted by the EPA (1983, 1986), Grizzard, et al. (1986), Randall et al. (1982), Rinella and McKenzie (1982), and Whipple and Hunter (1981). In addition, Peter Stahre studied particle size distribution in Sweden. Stahre used a light blocking instrument developed by Wiksell (1976) which can measure 15 different TSS size intervals ranging from 2 to 500 microns (μm). All of the samples analyzed had first undergone a coarse separation using five minutes of sedimentation.

Stahre's findings are shown as a bar graph in Figure 21.3. As can be seen in this figure, the number of particles is greatest in the 10 to 20 μm range, while the numbers for particles larger than 40 μm are less than 10 particles per milliliter. Because of their small numbers, they do not appear on this graph.

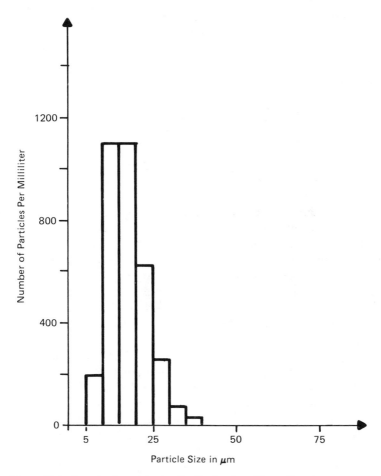

Figure 21.3   Sample of particle size distribution found in studies of stormwater by Stahre in Sweden.

A similar study, but using different instrumentation, was performed by Rinella and McKenzie (1982) of the U.S. Geological Survey in Portland, Oregon. They, on the other hand, did not use coarse separation of sediments before performing particle size analysis. We have summarized the results of their analysis of the February 12, 1981 Fanno Creek sample in Table 21.4. It can be seen that only 30% of all TSS by weight was found to have particles with settling velocities larger than 32 μm in diameter.

A similar study in Virginia by Randall et al. (1982) found that 80% of particles by weight were less than 25 μm, 89% were less than 35 μm, and 93% were less than 45 μm in size. The common theme among the three investigators supports the findings of Sartor and Boyd (1977) that most suspended pollutants in stormwater are associated with relatively small particles.

TABLE 21.4   TSS Distribution by Size of Particles in February 12, 1981
Sample of Fanno Creek

Particle Size (Microns)	TSS Load mg/l	Percent Finer than Given Size	Percent Increment of Sample
All sizes	832	100	100
<62	701	84	16
<31	584	70	14
<15	391	47	23
<8	262	31	16
<4	222	27	4

After Rinella and McKenzie, 1982

A word of caution about the data reported in this book. All of the data are representative only of the sampling sites from which they were taken. They may or may not be representative of other urban sites. Although we believe that the data reported here provide reasonable trends found in urban runoff, they should not be construed as applicable for any specific site you may study. We suggest that local data be collected for design and planning purposes.

21.2.6 Particle Volume Distribution

To help understand the makeup of the TSS in stormwater, we also discuss the distribution of volumes and surface areas by size of particles. A microscopic study of stormwater samples revealed that for the most part, particles in stormwater can be considered to be spherical in shape. This means that one can simply transform the particle size distribution into particle volume distribution. Figure 21.4 contains a bar graph of such transformation based on the same data used in the development of Figure 21.3.

An examination of Figure 21.4 reveals that very little volume of TSS is found in particles that are less than 5 microns in diameter. On the other hand, the 10 to 35 μm sized particles account for 90% of the TSS volume in this sample. Also of note is the fact that the volume of particles larger than 40 μm in diameter represent a significant fraction of the total TSS load. You may recall that this fraction did not even appear in Figure 21.3. This indicates that a relatively small number of particles can, in fact, represent a relatively significant load by volume and weight of the total TSS measured in stormwater.

As discussed earlier, some sites exhibit a first flush in pollutant loads. At those locations, the particle size and volume distribution can vary significantly throughout the storm. How the TSS volume distribution varied with time during one stormwater runoff event in Sweden is illustrated as a bar graph in Figure 21.5. This point is further illustrated in Figure 21.6 as a line graph of the total TSS volume vs. time from beginning of storm. From both of these figures, it is evident that the total particle volume in milliliters per liter de-

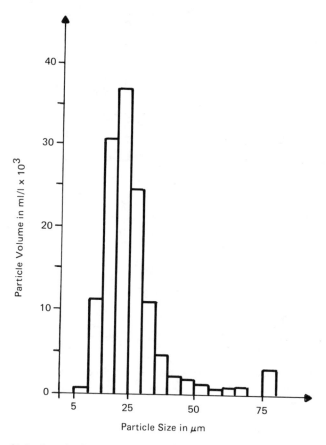

Figure 21.4   Sample of particle volume distribution found in studies of stormwater by Stahre in Sweden.

creased 75% in just 20 minutes. Such a finding at any location can significantly aid in the design of water quality detention facilities.

21.2.7 Particle Area Distribution

Knowledge of the distribution of particle volumes can indicate the weight of each particle size in TSS. Studies by Randall et al. (1982) revealed that the settling of pollutants occurs in two ways. First, the coarser particles settle out as discrete individual particles. The smaller particles, on the other hand, tend to agglomerate with time into larger particles, and their settling rate then accelerates. As a result, simple settling estimates may not reveal the actual mechanism of removal. The process can be multistepped and time-dependent. In other words, the removal of TSS and attached pollutants may in

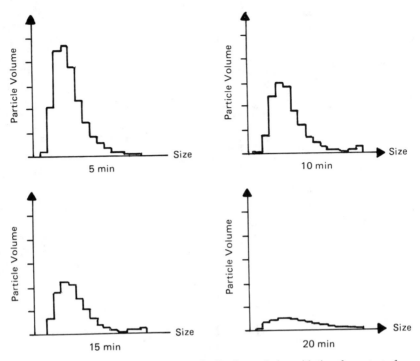

Figure 21.5   Sample particle volume distribution variation with time from start of runoff, according to Stahre.

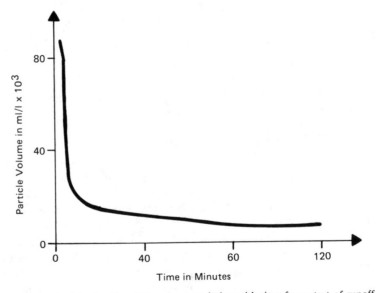

Figure 21.6   Sample of particle volume variation with time from start of runoff, according to Stahre.

fact occur in steps, and the smaller particles may not be removed if the residence time in the detention facility is insufficient for flocculation to take place. Knowledge of how the total surface area of the particulates varies with particle size can help us appreciate how other pollutants that attach to TSS may be removed. It is the available surface area that is used by ionic forms of other pollutants to bond to the particles. Also, agglomeration of smaller particles into larger ones is probably a function of the surface area available for the bonding to occur. A study by Grizzard et al. (1986) indicated that the majority of particle surface area was found in particles having a diameter of less than 50 to 60 microns (see Figure 21.7). These findings are based on the particle size distribution reported by Randall et al. (1982) and discussed near the end of section 21.2.5.

21.2.8 Density of Stormwater Pollutants

To calculate the settling velocity for a particle with a known size, it is necessary to know the density of that particle. There are other, more direct, ways of finding the particle settling velocity than by measuring particle sizes and their densities. These more direct techniques are described later, but for now we discuss particle density to help explain the basic principles of sedimentation in storage basins.

Density data is scarce and is often contradictory. Oscanyan (1975) reported 2,650 kilograms per cubic meter (i.e., specific gravity = 2.65) as the

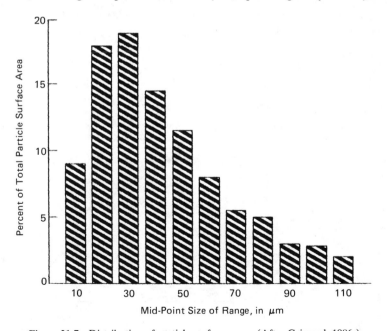

Figure 21.7   Distribution of particle surface areas. (After Grizzard, 1986.)

operative particle density. Obviously, this value is based on the specific gravity of solid rock materials and it is assumed that no pores are present within the TSS particles. Bondurant et al. (1975) reported TSS particle densities of 1,100 to 1,300 kilograms per cubic meter.

Peter Stahre performed a number of studies to determine the particle densities of TSS in stormwater samples collected in Sweden. His findings suggest that particle densities, among other things, depend on

- particle size,
- the pH of the water, and
- the content of heavy metals in the water.

Exactly how the density depends on these factors has not been clarified. It was found that it helps to describe the variations in densities if the particles are separated into two groups. The first is made up of particles having densities of 1,000 to 1,160 kilograms per cubic meter (light particles) and the second of particles having densities greater than 1,160 kilograms per cubic meter. For the latter, it was assumed that the average density of particles is 1,300 kilograms per cubic meter. For comparison, water has a density of 1,000 kilograms per cubic meter at a temperature of 20°C.

Figure 21.8 illustrates how the density distribution varies with particle size for the two density groups. We see that the proportion of heavier density particles declines with increasing particle size. At the same time, the proportion of lighter density particles increases with particle size. Also, some of the results suggested that the decrease for the heavier particles is less pronounced at lower pH, which may be related to dissolved metals in the water.

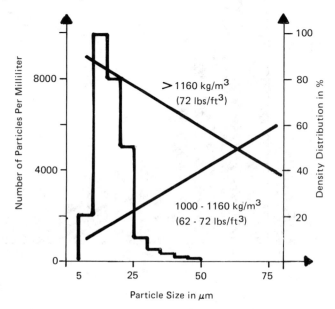

Figure 21.8   Sample TSS density distribution by particle size, according to Stahre.

21.2.9 Particle Settling Velocity Distribution

As mentioned earlier, settling velocity tests for samples of urban storm-water runoff were reported by EPA (1986), Rinella and McKenzie (1982), Randall et al. (1982), and Whipple and Hunter (1981). The most extensive set of data used for this purpose was in support of the EPA (1986) document. It was based on 50 different runoff samples from seven different sites, which among others included the data gathered under NURP and the results re-ported by Whipple and Hunter (1981).

The data was found to vary as much between runoff events at the same site as between different sites. As a result, all data were combined and pre-sented as typical for urban stormwater (see Figure 21.9). Based on this data and Figure 21.9, Table 21.5 was presented by the EPA (1986) which groups all the settling velocities as five distinct size fractions. While these "typical"

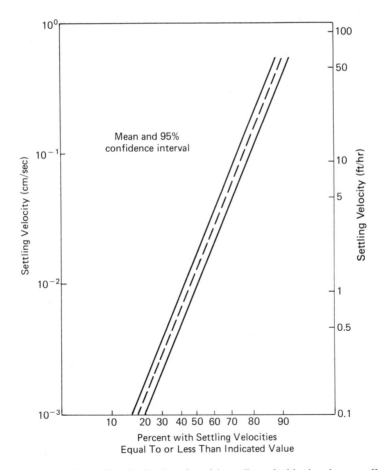

Figure 21.9 Probability distribution of particle settling velocities in urban runoff. (After EPA, 1986.)

TABLE 21.5   TSS Settling Velocity Distribution
for Five Size Fractions

Size Fraction	% of TSS in Urban Runoff	Average Settling Velocity (ft/hr)
1	0–20	0.03
2	20–40	0.30
3	40–60	1.50
4	60–80	7.00
5	80–100	65.00

After EPA, 1986

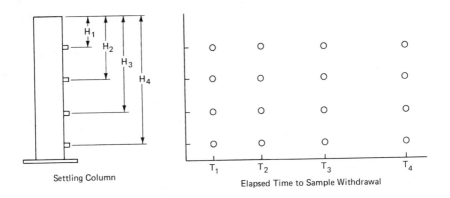

O = Data Point - Record % removed based on observed vs. initial concentration

Settling velocity (V_S) for that removal fraction is determined from the corresponding sample depth (h) and time (t)
$V_S = H/T$

Observed % removed reflects the fraction with velocities equal or greater than computed V_S

A probability plot of results from all samples describes the distribution of particle settling velocity in the sample

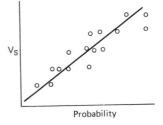

Figure 21.10   Particle settling velocity measurement equipment and procedure. (After EPA, 1986.)

results can suffice for initial planning estimates, it may be wise to obtain site-specific settling velocity data for final design.

Site-specific settling velocity measurements are not difficult to perform. The equipment for such measurements was suggested by Whipple (1981), Rinella and McKenzie (1982), and others. It is briefly described, along with the suggested procedure, in the 1986 EPA publication on detention basins. It basically consists of a vertical clear plastic settling cylinder approximately 6 inches in diameter and 6 feet high (see Figure 21.10). Usually, four sampling ports are installed in its side at 1-foot intervals measured from the top.

The tube is filled with stormwater runoff sample and small samples are withdrawn from the ports at preset time intervals. These are usually taken at 0.5, 1, 2, 4, 8, 12, 24, and 48 hours, and the samples are analyzed for TSS and/or other constituents. The constituent concentrations are then compared against the initial, fully-mixed sample, and the concentrations and the percent removal, port depth h, and time t are recorded. The settling velocity is simply the ratio of port depth and time (i.e., h/t). Thus, each percent removal measurement is plotted as the percent with a settling velocity equal to or less than the indicated value vs. settling velocity (see Figure 21.9). Figure 21.10 illustrates the equipment and the calculating procedure as it was presented in the EPA's 1986 publication titled *Methodology for Analysis of Detention Basins for Urban Runoff Quality.*

REFERENCES

BONDURANT, J. A., BROCKWAY, C. E., AND BROWN, M. J., "Some Aspects of Sedimentation Pond Design," *National Symposium on Urban Hydrology and Sediment Control,* University of Kentucky, 1975.

DRCOG, *Denver Urban Runoff Evaluation Program, Final Report,* Denver Regional Council of Governments, Denver, Colo., 1982.

EPA, *Results of the Nationwide Urban Runoff Program,* Final Report, U.S. Environmental Agency, NTIS Accession No. PB84-185552, December, 1983.

EPA, *Methodology for Analysis of Detention Basins for Control of Urban Runoff Quality,* U.S. Environmental Agency, EPA440/5-87-001, September, 1986.

GRIZZARD, T. J., RANDALL, C. W., WEAND, B. L., AND ELLIS, K. L., "Effectiveness of Extended Detention Ponds," *Urban Runoff Quality—Impacts and Quality Enhancement Technology,* Proceedings of an Engineering Foundation Conference, ASCE, 1986.

OSCANYAN, P., "Design of Sediment Basins for Construction Sites," *National Symposium on Urban Hydrology and Sediment Controls,* University of Kentucky, 1975.

RANDALL, C. W., ELLIS, K., GRIZZARD, T. J., AND KNOCKE, W. R., "Urban Runoff Pollutant Removal by Sedimentation," *Stormwater Detention Facilities,* Proceedings of an Engineering Foundation Conference, ASCE, 1982.

RINELLA J. F., AND MCKENZIE, S. W., "Determining the Settling of Suspended Chemicals," *Stormwater Detention Facilities,* Proceedings of an Engineering Foundation Conference, ASCE, 1982.

SARTOR, J. D., AND BOYD, G. B., "Water Pollutant Aspects of Street Surface Contaminants," U.S. Environmental Protection Agency, EPA-600/2-77-047, 1977.

STOCKHOLM'S WATER AND SEWER WORKS, *Stormwater Studies at Jarvafaltet,* 1978. (In Swedish)

WHIPPLE, W. JR., AND HUNTER, J. V., "Settling Ability of Urban Pollution," *Water Pollution Control Federation Journal,* Vol. 53(12), pp. 1726–31, December, 1981.

WIKSELL, H. "Development of an Instrument for Measuring Particle Size Distribution in Water and Sewerage Engineering," Vatten No. 1, 1976. (In Swedish)

22

Fundamentals of Sedimentation

22.1 INTRODUCTION

Sedimentation occurs when particles have a greater density than the surrounding liquid. Under laboratory quiescent conditions, it is possible to settle out very small particles; the smallest practical settling size in the field is around 10 micrometers (Metcalf & Eddy, 1979). The smallest particles have been observed sometimes to become electrically charged, which can further interfere with their ability to settle out. The fact is that we do not know if there is a particle size limit for separation by settling in water. If there is a lower limit, it probably is site-specific.

We briefly describe the following basic relationships that are often used to quantify the sedimentation process:

- Newton's formula,
- Stoke's law, and
- Hazen's surface load theory.

22.2 Newton's and Stoke's Sedimentation Laws

For spherical particles falling through a liquid, Newton suggested the following formula to define their maximum fall velocity:

$$v_s = \sqrt{\frac{4}{3} \cdot \frac{d \cdot g \cdot (r_p - r_v)}{C_D \cdot r_v}} \tag{22.1}$$

in which v_s = fall velocity of the particle,
    d = diameter of the particle,
    r_p = density of the particle,
    r_v = density of the fluid,
    g = acceleration of gravity, and
    C_D = drag coefficient of the particle.

The drag coefficient C_D will depend on whether the flow around the particle is laminar or turbulent and is a function of the Reynolds Number (Re) (see Figure 22.1). For Reynolds Numbers between 0.3 and 10,000, the drag coefficient can be approximated using the following equation:

$$C_D = \frac{24}{Re} + \frac{3}{\sqrt{Re}} + 0.34 \qquad (22.2)$$

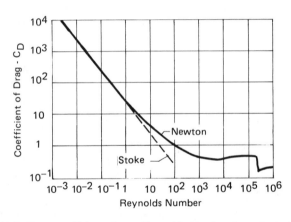

Figure 22.1 Drag coefficient C_D vs. Reynolds Number for spherical particles. (After McCabe and Smith, 1976.)

For Reynolds Numbers smaller than 0.3, one can neglect the last two terms of the preceding equation, and the drag coefficient can be approximated using

$$C_D = \frac{24}{Re} \qquad (22.3)$$

By substituting the following equation into the preceding expression

$$Re = \frac{v_s \cdot d \cdot r_v}{\mu} \qquad (22.4)$$

and combining it with Newton's Formula for particle fall velocity, one obtains Stoke's law:

$$v_s = d^2 \cdot g \cdot \frac{(r_p - r_v)}{18 \cdot \mu} \tag{22.5}$$

in which μ = dynamic viscosity of the fluid.

The fall velocity is directly proportional to the square of the particle diameter and the difference in the densities between the particle and the fluid. In water, Stoke's law is applicable to particles having an equivalent spherical diameter of up to 100 microns.

If the dynamic viscosity of the water and the density of the particles are known, the fall velocity can be calculated as a function of particle diameter. Figure 22.2 shows the results of such a calculation assuming the fluid is water at 15°C. As can be seen, the particle's density has a significant effect on the fall velocity. For example, a decrease in density from 2,000 to 1,500 kilograms per cubic meter will reduce the particle's fall velocity by approximately one-half.

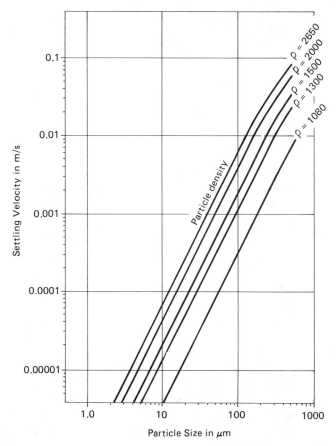

Figure 22.2   Theoretical fall velocity of spherical particles in water at a temperature of 15°C.

22.3 HAZEN'S SURFACE LOAD THEORY

Hazen's surface load theory assumes that for a particle to be permanently removed from the water column, it must reach the bottom of a basin before the water carrying it leaves the basin. Consider a long rectangular basin of length L, width W, and depth D. The surface area A of the basin is then

$$A = W \cdot L \tag{22.6}$$

and the volume V is

$$V = A \cdot D \tag{22.7}$$

Further, assume that fluid passing through the basin at a flow rate of Q is uniformly distributed over the cross-section $W \cdot H$ and that all the particles which have time to sink to the bottom will be permanently removed from the fluid. The descent height is the largest for particles entering the basin at the water surface (i.e., D, the depth of the basin).

The time T for the flow to pass through the basin can be given by

$$T = \frac{V}{Q} = \frac{(A \cdot D)}{Q} \tag{22.8}$$

For a particle to settle to the bottom as it passes through the basin, its average descent velocity has to be at least

$$v_s = \frac{D}{T} = \frac{Q}{A} \tag{22.9}$$

It can thus be stated that the sedimentation effect of a basin can be expressed by the ratio between Q and A, which is sometimes referred to as the surface load. Equation 22.9 states that the surface load is equal to the descent rate of the smallest particle that can just be separated in the basin.

This surface load theory presupposes that the flow through the basin is uniform and laminar. Unfortunately, these are not the conditions found in practice. A field installation can experience multilayered flow, turbulence, eddies, circulation currents, diffusion at inlets and outlets, etc. (see Figure 22.3). Some investigators speculate that under turbulent conditions no more than 60% of the removal predicted using Hazen theory is achieved. In design, correction factors are used to compensate for this observed difference between theory and actual performance.

According to the preceding equations, depth has nothing to do with sediment removal in a basin. However, because of turbulence, diffusion, and local velocities, sediments can be resuspended from the bottom. To reduce the chances of resuspension, it is recommended that the average basin depth be no less than 3.5 feet. It is suggested, however, that sedimentation basins be between 5 and 12 feet deep.

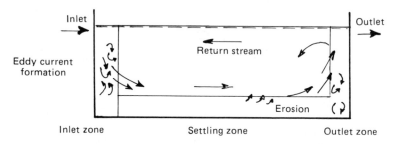

Figure 22.3   Examples of flow disturbances in a basin.

22.4 SEDIMENTATION IN STORMWATER UNDER QUIESCENT CONDITIONS

Until the early 1980s, very few studies dealt with the separation of pollutants from stormwater by sedimentation. One of these is further commented on here; namely, the work by Peter Stahre in Sweden. Since then, studies by Rinella and McKenzie (1983), Randall et al. (1982), and Whipple and Hunter (1981) have produced significant new information about the settling characteristics of TSS and associated pollutants. Nevertheless, the literature on this topic is still very limited.

22.4.1 Stahre's Findings

A more comprehensive study of sedimentation properties of TSS in stormwater was conducted by Stahre. He investigated how particle size distribution and particle volume varies with time. Using a pipette, Stahre sampled water in a settling tube at various times after settling was permitted to begin. During the first hour, samples were taken at 5-minute intervals. After that, additional samples were taken at 90 and 120 minutes. Each sample was analyzed for particle size distribution, and the particle volume distribution by particle size was calculated.

Figures 22.4 and 22.5 depict Stahre's findings of particle numbers and volumes for each size fraction in the water column as a function of sedimentation time. Both figures show results only for particles smaller than 25 microns.

As can be seen in Figure 22.4, the number of particles in the 5 to 10 micron size appears to increase rapidly during the first hour and continues to increase for a total of 90 minutes from the start of the test. After that, the number of particles appears to decrease. Something similar was observed for the 10 to 15 micron size fraction, except the numbers increased only very slightly for the first 15 minutes.

An explanation for this apparently unusual finding was offered by Stahre. He speculated that the equipment that counted the particles may have

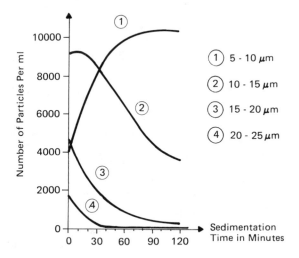

Figure 22.4 Number of particles in water column as a function of sedimentation time.

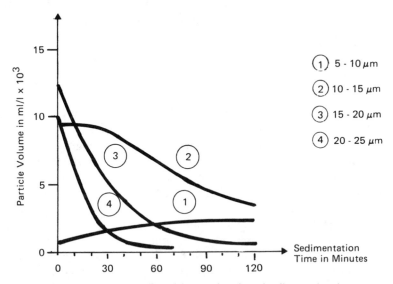

Figure 22.5 Volumes of particles as a function of sedimentation time.

mistaken small air bubbles in the water for particles. He did not, however, explain what may have caused the small air bubbles to appear early in the test. Regardless, because the 5 to 10 micron particles are very small, they contribute very little to the volume estimates of the suspended solids in the water column (see Figure 22.5).

By compositing all of the size fractions into a single volume of suspended solids, Stahre obtained a very smooth, exponentially decaying curve. This is shown in Figure 22.6, where the effect of sedimentation time is related to the total volume of suspended solid particles remaining in the water column.

22.4.2 Randall's Findings

Randall et al. (1982) reported results of laboratory settling tube tests of seven urban stormwater runoff samples. They found that the TSS concentration after 48 hours of sedimentation leveled off to between 5 and 10 milligrams per liter (see Figures 22.7 and 22.8). This is similar to the findings reported for sedimentation tunnels in Sweden, where the TSS concentrations bottomed out at 10 milligrams per liter. Although the settling tube tests appear to have somewhat lower final concentrations, both sets of results indicate a practical bottom limit of approximately 10 milligrams per liter in the removal of TSS by sedimentation.

The Randall et al. findings confirm another observation in Sweden, namely the percentage of TSS removed increased as the initial TSS concentration increased. Stahre and Urbonas plotted Randall's data as percent TSS

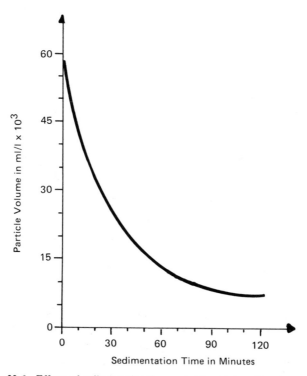

Figure 22.6   Effects of sedimentation time on total particle volume in stormwater.

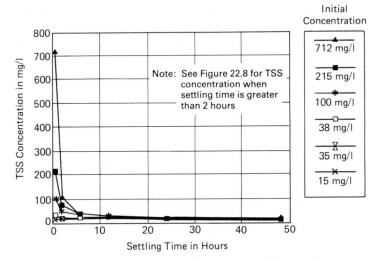

Figure 22.7   Effects of time of sedimentation on TSS concentrations. (After Randall et al., 1982.)

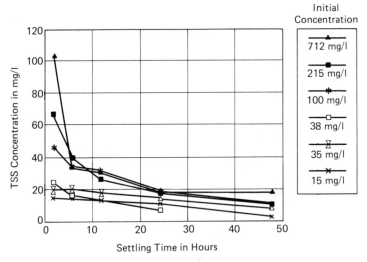

Figure 22.8   Effects of time of sedimentation, 2 to 48 hours, on TSS concentrations. (After Randall et al., 1982.)

removed vs. initial concentration (see Figure 22.9). This graph clearly indicates that the TSS removal efficiencies are very poor when initial concentrations are around 10 milligrams per liter. The removal efficiencies increase rapidly as the initial concentration increases to about 100 milligrams per liter, after which the removal efficiency begins to level off.

In addition to TSS, Randall's group also conducted settling characteristics tests in settling tubes for several other constituents found in the same stormwater samples. None of the other constituents exhibited the same consistency or uniformity in removal efficiencies found for TSS. However, it was clear that sedimentation was able to reduce their concentrations in water. Figures 22.10 through 22.14 contain graphs showing percent removal vs. sedimentation time for several of the constituents.

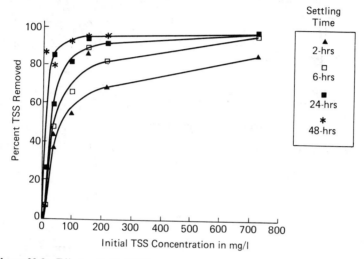

Figure 22.9 Effects of initial TSS concentration on removal rates. (After Randall et al., 1982.)

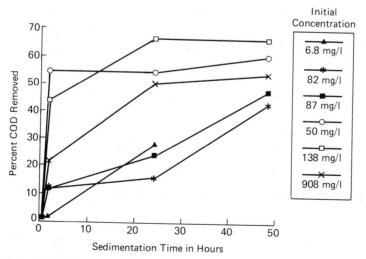

Figure 22.10 Percent COD removed vs. sedimentation time. (After Randall et al., 1982.)

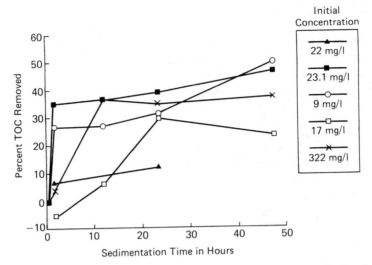

Figure 22.11   Percent TOC removed vs. sedimentation time. (After Randall et al., 1982.)

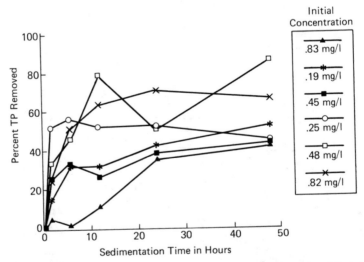

Figure 22.12   Percent TP removed vs. sedimentation time. (After Randall et al., 1982.)

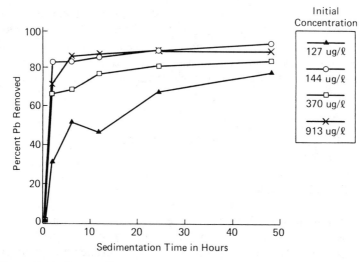

Figure 22.13   Percent Pb removed vs. sedimentation time. (After Randall et al., 1982.)

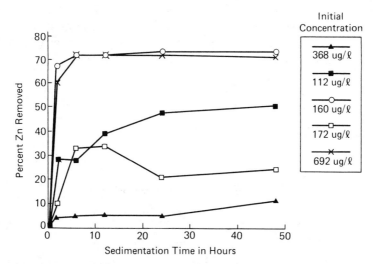

Figure 22.14   Percent Zn removed vs. sedimentation time. (After Randall et al., 1982.)

After 48 hours of sedimentation, COD removal was observed to range
between 40% and 65%, TOC between 20% and 50%, TP between 40% and
80%, Pb between 75% and 95%, and Zn between 10% and 70%. All of these
ranges can be viewed as possible maximum removal rates in detention ponds
which, because of their dynamic flow conditions, are not likely to be as
efficient.

22.5 SEDIMENTATION UNDER FLOW-THROUGH CONDITIONS

22.5.1 Kuo's Findings

Sedimentation in stormwater detention basins does not occur under
static conditions. Typically, sedimentation basins are designed for chronological flow-through. Unfortunately, very few experiments of sedimentation
under these conditions have been performed, with Kuo (1976) being one of
the few to have actually conducted such tests. He reported his results from
tests conducted in a flow-through setup depicted in Figure 22.15.

In the first set of tests, Kuo held the basin length constant at 6 meters
and the depth was varied between 1 and 4 meters. The tests were performed
using water carrying sediment having particle sizes of 55, 126, and 290 microns. Kuo's findings for the first set of these tests were plotted as the ratio of
basin length over depth (i.e., L/D) vs. percent sediment removed for different
particle sizes (see Figure 22.16). These figures indicate that the sediment

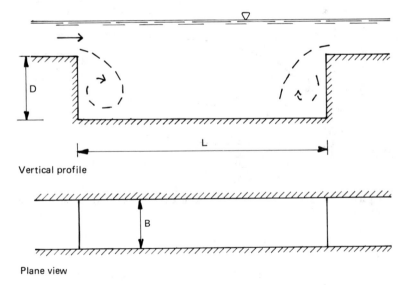

Vertical profile

Plane view

Figure 22.15   Test basin for flow-through separation of TSS. (After Kuo, 1976.)

removal efficiency is independent of basin depth, which supports the surface load theory derived by Hazen.

In the next set of experiments, Kuo held the basin depth constant at 2.4 meters and varied the length between 6 and 30 meters. Again, the basin L/D vs. percent sediment removed were plotted for different particle sizes (see Figure 22.17). As indicated in the figure, TSS removal improved with basin length. The improvement was most apparent for the larger sized particles.

Next, Kuo held the basin depth and basin length constant and varied the flow through rate between 1.13 m^3/s and 3.40 m^3/s. Figure 22.18 contains a plot of these results, showing that sedimentation improved as the flow rate, and the resultant surface loading, decreased. Again, the differences were most signifi-cant for the largest particles. Nevertheless, Kuo's findings appear to support

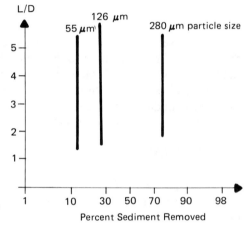

Figure 22.16   Sediment removal as affected by basin depth (for $L = 6$ m). (After Kuo, 1976.)

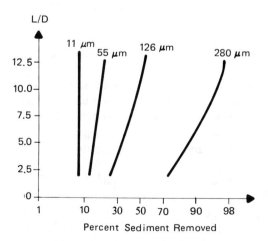

Figure 22.17   Sediment removal as affected by basin length (for $D = 2.4$ m). (After Kuo, 1976.)

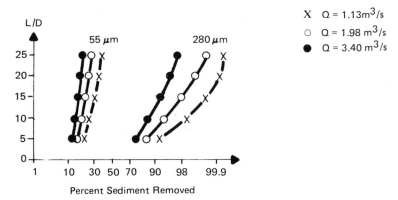

Figure 22.18   Sediment removal as affected by flow-through rate. (After Kuo, 1976.)

the original surface loading theory proposed by Hazen. We have to stress that these tests were performed under controlled laboratory conditions. In the field, the flows vary with time, there is a possibility of particle resuspension, and density and temperature currents are likely. As a result, we caution in using these findings and suggest that a safety factor be applied to compensate for unpredictable field conditions.

22.5.2 Grizzard's Findings

Grizzard et al. (1986) reported the results of a study comparing the laboratory settling tube test results against the pollutant removal rates at a prototype field installation. The facility being studied received runoff from a townhouse complex of 34.4 acres having 19.2% impervious cover. The extended detention basin was expected to drain completely between storms and was designed with a brim full water quality detention volume of 38,000 cubic feet. This is equivalent to 0.3 inches of runoff from the entire site or 1.55 inches of runoff from the impervious surfaces only.

The outlet was sized to drain the total water quality detention volume in 40 hours. As a result, the average drawdown time for all storm events during the test period was expected to be approximately 24 hours. Due to unanticipated leakage at the outlet, the test basin emptied in approximately 10 hours when full, and the more frequent storm events drained in approximately six hours.

The stormwater samples were compared to the laboratory settling tube results described in section 22.4.3. Grizzard found that for many of the pollutants, the prototype extended detention basin with a six-hour drawdown period had removal efficiencies comparable to those found in settling tubes after approximately two hours. There were several exceptions to this.

One was for total phosphorous (TP). The removal rate for TP in the field was 6 hours, compared to the one-half hour removal rate in the settling tube. In other words, the basin performed very poorly in removing TP when compared to settling tube tests.

Another exception was the results for the removals of Pb and Zn, for which the field removal rate was equivalent to the settling tube test results after 24 hours of settling. Thus, for at least these particular tests, the pond out-performed the settling tube results in removing Pb and Zn.

It is worth noting that the lesser removal efficiencies in the pond for TP are consistent with findings by others for the removal of TP in dry ponds. Outside of this, the observed trends in an extended detention dry pond compared favorably with the settling tube results. Thus, by applying a safety factor to tube settling results, an extended detention pond outlet design could be based on settling tube test findings.

REFERENCES

GRIZZARD, T. J., RANDALL, C. W., WEAND, B. L., AND ELLIS, K. L., "Effectiveness of Extended Detention Ponds," *Urban Runoff Quality—Impacts and Quality Enhancement Technology,* Proceedings of an Engineering Foundation Conference, ASCE, 1986.

KUO, C. Y., "Sedimentation Routing in an In-Stream Settling Basin," *Proceedings of the National Symposium on Urban Hydrology, Hydraulics and Sediment Control,* University of Kentucky, 1976.

LISPER, P., *Pollution in Storm Water and Its Variations,* Dissertation, CTH, Sweden, 1974. (In Swedish)

MCCABE, W. L., AND SMITH, J. C., "Unit Operations of Chemical Engineering," 3rd edition, McGraw-Hill, 1979.

METCALF & EDDY, INC., *Wastewater Engineering, Treatment, Disposal, and Reuse,* 2nd edition, McGraw-Hill, 1976.

RANDALL, C. W., ELLIS, K., GRIZZARD, T. J., AND KNOCKE, W. R., "Urban Runoff Pollutant Removal by Sedimentation," *Proceedings of the Conference on Stormwater Detention Facilities,* ASCE, 1982.

RINELLA, J. F., AND MCKENZIE, S. W., *Methods for Relating Suspended-Chemical Concentrations to Suspended-Sediment Particle Size Classes in Storm-Water Runoff,* U.S. Geological Survey Water-Resources Investigation 82-39, Portland, Oreg., 1983.

WHIPPLE, W., AND HUNTER, J. V., "Settleability of Urban Runoff Pollution," *Journal Water Pollution Control Federation,* Vol. 53, No. 12, 1981, pp. 1726–31.

23

Design of Water Quality Basins for Stormwater

23.1 INTRODUCTION

It is clear that the design of stormwater quality enhancement basins is an emerging technology. We attempted to provide in this book some of the technical basis for the removal of pollutants by sedimentation, which appears to be the primary water cleansing mechanism. Apparently, there are other water quality mechanisms at work also. As an example, in properly sized basins with a permanent pool of water, there is evidence of biological activity at work in the removal of nutrients (Randall, 1982).

The engineering profession is only beginning to learn about this technology, and the best we can offer here is what we have learned so far. More important, we caution you not to rely fully on any technique presented here or in any other publication to date. As in any emerging technology, there are many questions to be answered before we can design stormwater quality enhancement basins with complete confidence in how they will perform in the field. Also, we are dealing with random processes for which we can, at best, design only for a certain range of probable events. Larger storm events will happen. It is only a question of how often and when.

We discuss several design approaches, ranging from very technical to totally empirical and rule of thumb. We do not pass judgment on which of these basin sizing methods will give the best results. That will depend on the designer and on the water quality goals. All of the sizing methods are for the removal of the fine sediments and the constituents associated with fine sediments. In addition, we suggest a procedure for how to remove coarse sediments.

23.2 SIZING OF BASINS USING SURFACE LOAD THEORY

23.2.1 Basic Relationships

Separation of sediments from the water column occurs within the body of a detention pond. You may recall that according to the surface load theory, water depth in a pond should have no role in this separation process. We know, however, that this theory is only valid under uniform, steady, laminar flow conditions, which are not found in stormwater detention ponds.

To help us estimate sedimentation under turbulent flow conditions, we turn to the works of Dobbin (1944) and Camp (1946). They derived an analytical expression for sedimentation under turbulent flow conditions. Figure 23.1 is based on their work and shows how three nondimensional parameters relate, namely:

$$\frac{V_s}{\dfrac{Q}{A}}, \frac{V_s H}{2\epsilon}, \text{ and } E \qquad (23.1)$$

in which V_s = settling velocity of a particle,
Q/A = surface load,
H = water depth in the basin,
ϵ = diffusion coefficient, and
E = sedimentation effectiveness (i.e., the fraction of sediment removed).

With the help of Figure 23.1, it is possible to estimate the sedimentation effectiveness for various particle sizes. Examining this graph, we see that sedimentation is influenced by the flow conditions in the basin. Small values of $V_s H/2\epsilon$ correspond to turbulent flow and large values represent laminar flow conditions. As the value for this parameter increases, sedimentation effectiveness E approaches the value

$$E = \frac{V_s}{\dfrac{Q}{A}} \qquad (23.2)$$

which is in full agreement with Hazen's surface load theory. Now, if we assume

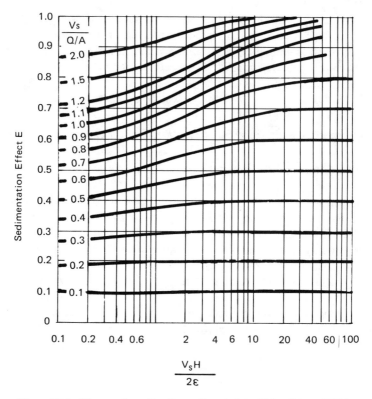

Figure 23.1 Diagram for estimating sedimentation. (After Camp, 1946.)

that the velocity distribution in the basin is parabolic, the diffusion coefficient, according to Camp (1946) can be expressed as

$$\epsilon = 0.075H \cdot \sqrt{\frac{\tau}{\rho}}$$ (23.3)

in which H = water depth,
  τ = critical shear stress, and
  ρ = water density.

The value of critical shear stress is determined by the size and density of the smallest particle that will not be moved by water flowing through the basin. Figure 23.2 relates the values of critical shear stress for silicon particles to particle size and the purity of the water. This figure represents the recommendations by U.S. Bureau of Reclamation (1973) for non-eroding shear stress in channels. Notice that the critical shear stress stops declining when the particle sizes become small. This is attributed to cohesive forces between the smallest particles. Thus, if we use the critical shear stress values found at the extreme left of Figure 23.1, we should be working in a fairly conservative and stable range.

If the settling velocities, V_s, are known for all particles in the stormwater passing through a pond, we can estimate the sedimentation effectiveness of each size group with the aid of Figure 23.1. This can be done for any known or desired design surface loading (i.e., Q/A). Figure 23.3 presents the results of such calculations under the assumption that the parameter $V_sH/2\epsilon$ is equal to 0.1 (i.e., turbulent flow) and that only gravitational and tractive forces are at work during sedimentation. This assumption is questionable for particles smaller than 5 to 10 microns in diameter, because electrostatic forces can influence their settling rates.

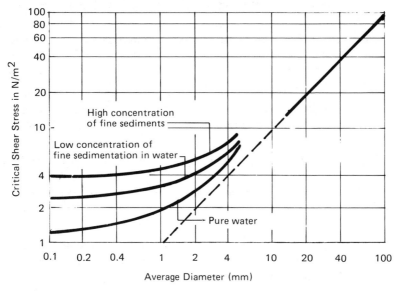

Figure 23.2 Critical bed load shear stress in channels. (After U.S. Bureau of Reclamation, 1973.)

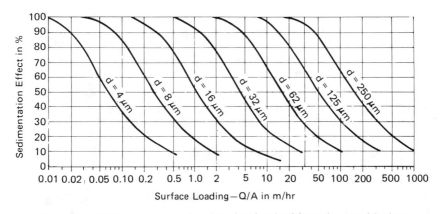

Figure 23.3 Sedimentation as a function of surface load for various particle sizes.

23.2.2 Suggested Pond Sizing Procedure

The preceding information can be put together into a procedure which can be used to size water quality enhancement ponds. The following suggested procedure has a physical basis and a theoretical foundation:

1. Determine the particle size and particle volume distribution associated with the pollutants in your stormwater samples.
2. Decide on the basis of the sediment data how much of the various particle sizes will need to be removed to achieve the desired water quality. Keep in mind that in some cases sedimentation alone may not achieve your goals.
3. Using Figure 23.3, make a preliminary estimate of the maximum allowable hydraulic surface loading rate.
4. With the aid of Figure 23.2, estimate the settling velocities for all representative particle sizes.
5. Using the information developed in steps 1 through 4 and in Figure 23.1, calculate the sedimentation effectiveness for all of the representative particle sizes.
6. Composite the total sedimentation effectiveness using the results from step 5.
7. If the composite total sedimentation effectiveness is too small, repeat the calculations using a lower hydraulic surface loading rate than assumed in step 3.
8. After you are satisfied with the adequacy of the hydraulic surface loading rate, determine the surface area of the pond and its configuration. The hydraulic surface loading rate used can be based on the average flow-through rate in the storage basin.

The preceding calculating procedure is far from being fully developed. It is a procedure that can easily be converted into a computer algorithm and, with proper data and calibration, it could provide sound pollutant removal estimates in sedimentation ponds. How accurate these estimates will turn out to be is yet to be tested. We suspect that if the pond is properly configured geometrically, has good energy dissipation at the inlet, and the sediment/pollutant relationships for the stormwater are well-defined and the model is calibrated, this procedure should produce very good results.

23.3 SIZING OF PONDS USING 1986 EPA RECOMMENDATIONS

In 1986, the Environmental Protection Agency published a suggested methodology for the analysis of detention basins for the control of urban runoff pollution. This document, according to EPA (1986), provides analysis methodology to "... guide planning level evaluation and design decisions...." for the sizing of detention ponds. The procedure is an adaptation of probabilistic methodology originally formulated by DiToro and Small (1979) under partial funding by the EPA and reported in a Hydroscience, Inc, (1979) report to the EPA. The actual methodology in the 1986 EPA publication was the work of Eugene D. Driscoll, with technical consultation from Dominic M. DiToro. All of these efforts were undertaken with Dennis Athayde of EPA as the Project Officer.

23.3.1 Relationships for Dynamic Conditions

The EPA (1986) methodology combines probabilistic techniques for the analysis of rainfall and runoff with the sedimentation theory for removal of sediments under quiescent and dynamic conditions. The approach used to estimate sediment removal under dynamic conditions is very similar to what was described in section 23.2 and is based on a similar sedimentation removal equation found in Fair and Geyer (1954), namely:

$$R_d = 1.0 - \left[1.0 + \frac{1}{n} \cdot \frac{V_s}{\dfrac{Q}{A}} \right]^{-n} \tag{23.4}$$

in which R_d = fraction of the initial solids removed under dynamic conditions,

V_s = settling velocity of particles,

Q = peak flow-through rate,

A = surface area of the detention pond, and

n = turbulence or short circuiting constant that is used to indicate the settling performance of the pond. Suggested values of n by Fair and Geyer (1954):

$n = 1$, poor performance,

$n = 3$, good performance,

$n > 5$, very good performance, and

n = infinity, ideal performance.

When n approaches infinity, Equation 23.4 reduces to the following, and familiar, exponential decay form:

$$R_d = 1.0 - e^{-kt} \tag{23.5}$$

in which $k = v_s/h$, a sedimentation rate coefficient,

h = average depth of the basin,

$t = V/Q$, residence time, and

V = volume of basin.

Solving Equations 23.4 and 23.5 and plotting the results gives us a graph that looks very much like the one in Figure 23.1. Reducing these calculations further to a range of particle settling velocities and surface loading rates (i.e., flow-through rates) most probable under urban runoff conditions results in a graph shown in Figures 23.4 and 23.4a. Figure 23.4 is the exact solution of Equation 23.4, assuming $n = 3$ (i.e., good removal conditions), and Figure 23.4a is the exact solution assuming $n = 1$ (i.e., poor conditions). The latter of the two may be the more appropriate solution under dynamic field conditions of variable inflow and depth.

The removal under dynamic conditions varies with storm intensity. As a result, estimates made for average seasonal or annual conditions have to be corrected to estimate the actual fraction of TSS removed over an extended period of time.

Long-term average removal is estimated with the aid of Equation 23.6. It accounts for the variations in hourly storm depths, and its graphic solution is presented in Figure 23.5. While the estimates made using average rainfall intensities and volumes according to EPA (1986) are not exact, they may be useful for planning level purposes. A more exact estimate of long-term removal rates can probably be obtained using similar principles with a chronological continuous model.

$$R_L = Z \cdot \left[\frac{r}{r - \ln\left(\dfrac{R_M}{Z}\right)} \right]^{(r+1)} \tag{23.6}$$

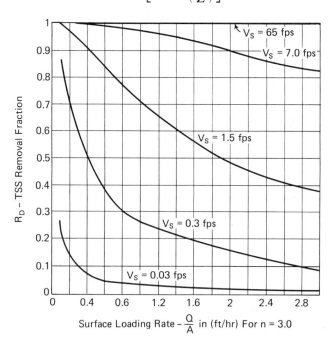

Figure 23.4   Loading rate vs. removal by sedimentation ($n = 3$).

in which R_L = long-term dynamic removal fraction,
$\quad R_M$ = mean storm dynamic removal fraction,
$\quad r = 1/CV_Q^2$,
$\quad CV_Q$ = coefficient of variation of runoff flow rates, and
$\quad Z$ = maximum fraction removed at very low flow rates.

When using Equation 23.6 or Figure 23.5, the value for Z can be assumed to be 80% to 100% for all fractions. An exception to this would be where the smallest sizes of sediments are electrostatically charged (i.e., clays), in which cases 10% to 50% would be more appropriate to use for the sediment fractions having 0.3 and 0.03 feet per hour settling velocities.

23.3.2 Relationships for Quiescent Conditions

Surface storm runoff occurs only during a fraction of the time in any given year. As a result, a significant amount of runoff is trapped and retained inside a pond where sedimentation continues between storms under quiescent conditions. How large the pond volume is relative to storm runoff volume determines the TSS removal effectiveness under quiescent conditions. The removal of TSS under quiescent conditions can be approximated by:

$$R_Q = V_S \cdot A_B \qquad (23.7)$$

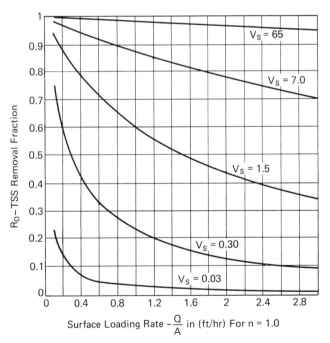

Figure 23.4a   Loading rate vs. removal by sedimentation ($n = 1$).

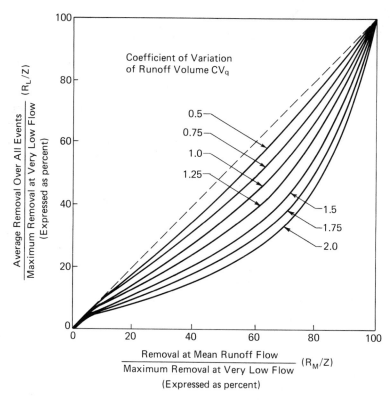

Figure 23.5 Long-term performance under dynamic conditions when removal is sensitive to flow rate. (After EPA, 1986.)

in which R_Q = solids removal rate, quiescent conditions,
V_S = settling velocity of a particle, ft/hr, and
A_B = pond surface area, ft².

When and for how long the quiescent periods occur is a random process. Although estimates of TSS removal can be made using long-term averages, these estimates need to be adjusted to reflect this random process. Such adjustments can be made with the aid of Figure 23.6.

To calculate the TSS removal using Figure 23.5, it is first necessary to find the ratio of *effective storage volume* to *mean runoff volume. Effective storage volume* is not the same as the total pond volume and is dependent on how and when rainstorms occur in relation to the quiescent periods.

The *effective storage volume* vs. *mean basin volume* ratio can be estimated using Figure 23.7. Its use, however, requires that we know the watershed's *mean runoff volume,* which can be found by using the nearest National Weather Service hourly precipitation records and processing them through a rainfall runoff model. Once we have all the necessary data, we enter the graph

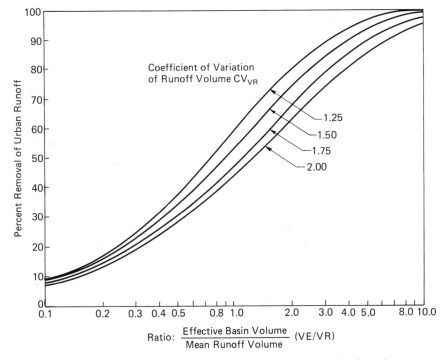

Figure 23.6   Average long-term performance of a detention pond under quiescent conditions. (After EPA, 1986.)

in Figure 23.7 with a value for the *storage basin volume* to *mean runoff volume* ratio and with a value for *volume of TSS removed under quiescent conditions* to *runoff volume* ratio as given by the following expression:

$$E = \frac{U_M \cdot R_Q}{V_R} \tag{23.8}$$

in which E = removal ratio under quiescent conditions,
   U_M = mean interval between storms, hours, and
   V_R = mean storm runoff volume for study period, ft^3.

The procedure and the examples published by EPA (1986) utilize rainstorm intensities and volumes averaged over a long data collection period. The use of monthly rainfall/runoff statistics and month-to-month analysis for an extended period of time should result in different TSS removal performance conclusions. Although long-term averages and variance statistics can be used for a quick and easy evaluation of how various facilities may remove TSS over an extended period of time, personal computers make it just as easy to perform analysis on a chronological series of data. This type of analysis should provide a better understanding of what to expect at any given site.

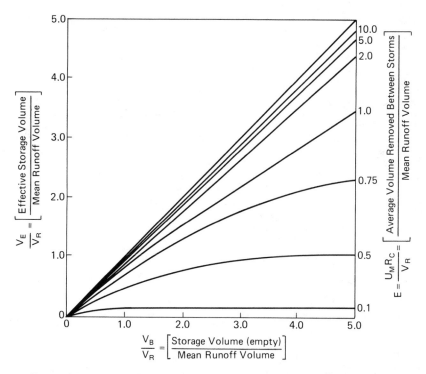

Figure 23.7   Effect of previous storms on long-term performance under quiescent conditions. (After EPA, 1986.)

23.3.3 Relationships for Combined Conditions

The relationships described so far can be used to estimate the removal to TSS under dynamic or quiescent conditions. Now let's combine both as suggested by EPA (1986) to estimate the total TSS removal by a detention pond.

If we define the fraction of time the pond is expected to operate under dynamic conditions as

$$f_D = \frac{D}{U_M} \qquad (23.9)$$

then the fraction of time under quiescent conditions is

$$f_Q = (1.0 - f_D) \qquad (23.10)$$

in which D = mean storm duration, and
U_M = mean interval between storm mid-points.

The TSS entering the pond first undergoes removal under dynamic conditions. For large ponds, this can occur more than one time since runoff from other storms is likely to occur before the inflow from an earlier storm fully drains from the pond. For a single dynamic removal period the removal efficiency is a function of the overflow rate divided by the effective pond surface area (i.e., loading rate).

The quiescent removal process operates on that fraction of inflow that remains in the pond after storm runoff ceases and the storm surcharge is emptied. The settling process continues on that fraction of TSS that was not removed during the dynamic conditions. The combined total fraction removed under both processes can be expressed as

$$f_T = 1.0 - F_D \cdot F_Q \qquad (23.11)$$

in which f_T = total TSS fraction removed,

F_D = fraction *not* removed under dynamic conditions, and

F_Q = fraction *not* removed under quiescent conditions.

23.3.4 Example 1: Long-term Average TSS Removal

Given: A 100-acre single family residential development in Arkansas has a runoff coefficient $C_V = 0.25$. A pond is built to intercept all runoff and has the following dimensions:

Surface area: $A_B = 12,000$ square feet
Average depth: $h_B = 4$ feet
Storage volume: $V_B = 48,000$ cubic feet

The rainfall statistics, as reported by the EPA (1986), are as follows:

	Units	COEFFICIENT OF VARIANCE	
		Mean	CV
Volume (V)	inch	0.52	1.54
Intensity (I)	in/hr	0.122	1.35
Duration (D)	hr	5.2	1.29
Interval between storms (U_m)	hr	87.	1.06

For this example, we assume that the particle settling velocities are the same as given in Table 21.5.

Required: Estimate the potential long-term removal rate of TSS from stormwater by the given detention pond.

Solution:

1. Calculate the mean storm runoff parameters:

$$\text{Flow rate: } Q_R = I \cdot C_v \cdot A = \frac{0.122 \cdot 0.25 \cdot 100 \cdot 43{,}560}{12}$$
$$= 11{,}100 \text{ cubic feet/hour}$$
$$\text{Volume: } V_R = V \cdot C_v \cdot A = \frac{0.52 \cdot 0.25 \cdot 43{,}560}{12}$$
$$= 47{,}200 \text{ cubic feet}$$

Assume that the runoff has the same statistical distribution as rainfall, namely,

$$CV_Q = 1.54 \quad \text{and} \quad CV_v = 1.35$$

2. Calculate removal under dynamic conditions:
The average loading rate during the mean storm is

$$Q_R/A_s = \frac{(11{,}100 \text{ ft}^3/\text{hr})}{(12{,}000 \text{ ft}^2)}$$
$$= 0.925 \text{ ft/hr}$$

Each size fraction of TSS will have its own settling velocity for which the removal by sedimentation, R_D, is calculated using Equation 23.4 assuming $n = 1$ (i.e., poor removal conditions), or Figure 23.4a.

Next, we need to find a value for Z (i.e., TSS removal at very low loading rates). This can be very subjective; however, assuming $Z = 1.0$ introduces very little error.

Using $Z = 1.0$ and R_D calculated with the aid of Figure 23.4a, we next calculate the long-term average removal for each size fraction with the aid of Figure 23.5 at $CV_Q = 1.54$.

The results for long-term dynamic removal are:

Size Fraction	V_S (ft/hr)	$(R_D/100)$ (Fig. 23.4)	Fraction Removed (Fig. 23.5)
1	0.03	.03	.03
2	0.3	.27	.12
3	1.5	.66	.36
4	7.0	.89	.70
5	65	.99	.99
	Total average fraction removed = 0.44		
	Fraction not removed $F_D = 1.0 - 0.44 = 0.56$		

3. Calculate removal under quiescent conditions:

Ratio of pond volume to mean runoff volume:

$$\frac{V_B}{V_R} = \frac{48,000}{47,200} = 1.02$$

We find the long-term removal efficiency under quiescent conditions with the aid of Figure 23.6, which relates the percent of TSS removal to the ratio of *effective basin volume/mean runoff volume* (i.e., V_E/V_R) and the coefficient of variation in runoff volume, $CV_V = 1.35$ defined in step 1.

The V_E/V_R ratio is not the same as the V_B/V_R calculated previously; instead, it is found using Figure 23.7. To do this, we need to first calculate E using Equation 23.8, namely:

$$E = \frac{U_M \cdot R_Q}{V_R} \quad \text{or} \quad E = \frac{U_M \cdot (A_B \cdot V_S)}{V_R}$$

in which E = removal ratio under quiescent conditions,
$R_Q = A_B \cdot V_S$, solids removal rate, ft^3,
V_R = volume of runoff, ft^3,
A_B = pond surface area, ft^2,
V_S = particle settling velocity, ft/hr, and
U_M = mean interval between storms, hr.

Thus, using all of the foregoing, we calculate the removal of each particle size fraction separately and find the total average fraction removed under quiescent conditions.

This yields the following results:

Size Fract.	V_S (f/s)	$A_B \cdot V_S$ (ft^3)	E	V_E/V_R (Fig. 23.7)	% Removed (Fig. 23.6)
1	0.03	360	0.7	0.70	47
2	0.3	3,600	6.7	0.95	57
3	1.5	18,000	33.2	1.02	58
4	7	84,000	155	1.02	58
5	65	780,000	1440	1.02	58

Total average removed = 56

$$\text{Fraction not removed } F_Q = \frac{100 - 56}{100} = 0.44$$

4. Calculate removal under combined conditions:

The combined total average long-term TSS removal is calculated using the values for F_D and F_Q calculated previously and Equation 23.10, namely:

$$f_T = (1 - F_D \cdot F_Q) = (1 - 0.56 \cdot 0.44) = 0.75$$

The combined total annual removal breaks down as follows:

Size Fract.	V_s (f/s)	% Dynamic Removal	% Quiescent Removal	% Combined Removal
1	0.03	3	47	48
2	0.3	12	57	62
3	1.5	36	58	73
4	7	70	58	87
5	65	99	58	100
Average:		44	56	75

You will notice that the quiescent removal rates for size fractions 4 and 5 are less than for dynamic removal. Because of the basin size, the statistical distribution of storms, and the intervals between storms, most of the removal process took place during the dynamic periods for these two fractions. In other words, quiescent removal hardly came into play.

5. All of these calculations assume that all TSS removal occurs because of sedimentation. These calculations do not account for any resuspension of sediments by inflow or wave action, remobilization of pollutants by chemical or biological processes in the pond, or the removal of pollutants by biological processes.

23.3.5 Example 2: Long-term Average Phosphorus Removal

Given: Use the same site, rainfall, runoff, and detention pond as described in the preceding example. This time we are given the following data concerning total phosphorus (i.e., TP) found in stormwater:

Total phosphorus: 0.71 milligrams per liter
Dissolved phosphorus: 0.32 milligrams per liter
Suspended phosphorus: 0.39 milligrams per liter

The suspended fraction has the following settling velocity distribution:

Size Fraction	V_S (ft/hr)	% of Mass in Urban Runoff
1	0.15	0–20
2	1.3	20–40
3	6.3	40–60
4	26	60–80
5	115	80–100

Required: The long-term removal rate of TP by the pond.

Solution:

1. Use the same storm runoff parameters found in Example 1, namely:

   Flow rate: $Q_R = 11,100$ cubic feet per hour
   $CV_Q = 1.54$

   Volume: $V_R = 47,200$ cubic feet
   $CV_v = 1.35$

2. Calculate removal under dynamic conditions:

   From Example 1, we have $Q_R/A_B = 0.925$ ft/hr.

   Again we assume $Z = 1.0$, and using Figure 23.4a we calculate the removal rate under dynamic conditions for each of the size fractions. Also, the long-term average removal for each size fraction is found with the aid of Figure 23.5 at $CV_Q = 1.54$. The results are summarized in the following table:

Size Fraction	V_s (ft/hr)	$(R_D/100)$ (Fig. 23.4)	Fraction Removed (Fig. 23.5)
1	0.15	0.15	0.08
2	1.3	0.50	0.23
3	6.3	0.85	0.63
4	26	0.95	0.95
5	115	1.00	1.00

Average fraction removed = 0.58

Fraction not removed $F_D = 1.0 - 0.58 = 0.42$

3. Calculate removal under quiescent conditions:

   From Example 1 we know that $V_B/V_R = 1.02$, $CV_V = 1.35$, and $U_M = 87$ hrs. We can now calculate the new values for E using

$$E = \frac{U_M \cdot (A_B \cdot V_S)}{V_R}$$

   and Figures 23.6 and 23.7. The results are as follows:

Size Fract.	V_S (f/s)	$A_B \cdot V_S$ (ft³)	E	V_E/V_R (Fig. 23.7)	% Removed (Fig. 23.6)
1	0.15	1,800	3.3	0.90	55
2	1.3	15,600	29	1.02	58
3	6.3	75,600	139	1.02	58
4	26	312,000	575	1.02	58
5	115	1,380,000	2540	1.02	58

Total average removed = 57

Fraction not removed $F_Q = \dfrac{100 - 57}{100} = 0.43$

4. Calculate removal under combined conditions:

The combined total average long-term removal of suspended phosphorus (SP) is calculated using the values for F_D and F_Q calculated in steps 2 and 3 and Equation 23.10, namely:

$$f_{SP} = (1 - F_D \cdot F_Q) = (1 - 0.42 \cdot 0.43) = 0.82$$

The combined total SP removal breaks down as follows:

Size Fract.	Vs (f/s)	% Dynamic Removal	% Quiescent Removal	% Combined Removal
1	0.15	8	55	59
2	1.30	23	58	68
3	6.30	63	58	84
4	26.00	95	58	98
5	115.00	100	58	100
Total averages:		58	57	82

The removal of total phosphorus is then found by compositing the dissolved phosphorus and the remaining suspended phosphorus. Thus, the average long-term TP concentration in the water leaving the pond is

$$TP = DP + (1 - f_{SP}) \cdot SP$$
$$= 0.32 + (1 - 0.82) \cdot 0.39 = 0.39 \text{ mg/l}$$

Thus, the long-term removal rate for TP is

$$F_{TP} = 100 \cdot \frac{0.71 - 0.39}{0.71}, \text{ or } 45\%$$

5. Note that despite an 82% removal rate for the suspended fraction, the removal of total phosphorus was less than 50%. This is because we assumed that no removal of the dissolved fraction takes place due to sedimentation. Also, as in the first example, these calculations do not account for any resuspension of sediments by inflow or wave action, remobilization of phosphorus by chemical or biological processes in the pond, or removal of dissolved phosphorus by biological processes.

23.3.6 Modification to the 1986 EPA Procedure

First, the technique which was just described is the same as recommended in EPA (1986). It assumes that the average hourly rainfall intensities can be used to adequately describe the sedimentation process and the long-term performance of the detention basins. This is not always the case. These assumptions may be reasonable if the storms are of low intensity type that last a long time. They may not be valid in the case of thunderstorms.

Thunderstorms can have very intense rainfall rates and be much shorter in duration than one hour. As a result, the average inflow rate throughout a thunderstorm can be much greater than estimated using hourly averages.

We suggest that the planner or the designer at least double the average hourly surface loading rate (i.e., Q/A) to perform the preceding calculations. Use of this safety factor is recommended until this technology is further verified by additional field observation. Until then, it is safer to err on the safe side.

Second, the EPA (1986) presented the results of a long-term rainfall statistical analysis for several regions of the United States. Although not completely clear, it appears that this analysis utilized full rainfall records as recorded by the National Weather Service. Apparently no effort was made to adjust the rainstorm population for probable runoff. The rainfall statistical information reported by EPA (1986) is included in Figure 23.8.

On examination of U.S. Weather Service rainfall records, we find that they contain many storms that will not produce surface runoff. If we apply basic hydrologic principles, we know that most rainstorms having 0.05 to 0.1 inches, or less, of rainfall will have no measurable runoff from paved surfaces, and rainstorms having 0.25 to 2.5 inches of rainfall will have no measurable runoff from unpaved surfaces, depending on soil, antecedent precipitation, and vegetative cover. It is therefore recommended that the rainfall statistics reported by EPA (1986) be used with this in mind. Namely, be aware that the

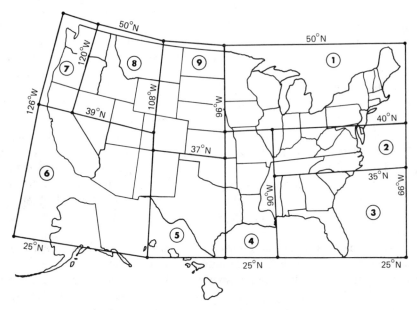

Figure 23.8 Regional rainfall statistics for preliminary estimates. (After EPA, 1986.)

rainfall averages reported in EPA (1986) may be low for estimating representative runoff statistics.

To assure yourself that the runoff statistics are more representative, we recommend that the local rainfall data be processed through a rainfall abstractions model. Statistical analysis of the resultant runoff volumes will provide the needed runoff volume averages, CV_vs, CV_Qs, durations, and periods between runoff events.

Third, more representative results are likely if continuous chronological runoff modeling and sedimentation removal are used. This can be done by transforming the hourly rainfall record into a runoff record using a simple rainfall–runoff model. This then can be processed through a chronological series of sediment removal calculations for the dynamic and quiescent periods, and a more representative picture of sediment removal can be developed. Such a model should be simple to write using any of the commercially available personal computer programing languages.

23.4 SIZING DRY BASINS

The design of dry detention basins for water quality is much less scientific than for the design of ponds. Although sedimentation is still the primary pollutant removal mechanism, the calculation of how much occurs has to be based mostly on empirical findings. Unlike a pond with a permanent pool of water, the dry basin fills during the storm and empties completely through an outlet at the bottom of the basin. As a result, surface load theory does not apply throughout the entire filling and emptying cycle. The water surface area changes throughout the storm, and the settling particles flow out through the outlet at the bottom instead of being trapped below the overflow. Also, no part of the runoff remains in the pond after it empties for sedimentation to take place under quiescent conditions.

Despite all of these apparent disadvantages, extended detention water quality basins can provide good removal rates for total suspended solids, lead, and zinc and fair removals for total organic carbon and chemical oxygen demand. Grizzard et al. (1986), Metropolitan Washington Council of Governments (1983), and Occoquan Watershed Monitoring Laboratory (1986) reported similar findings for the removal efficiencies of extended detention basins. From their findings, certain conclusions can be made concerning how to size these basins.

According to the results reported by Grizzard et al. (1986), the average storm runoff volume needs to be detained 24 hours in the extended detention basin to achieve equivalent total suspended solids removed after 6 hours in a settling tube. They suggested that the water quality basin volume be larger than the volume of runoff from an average rainstorm and that the outlet be designed so that the basin's full volume is drained in approximately 40 hours.

When designing an extended detention basin, the water quality basin volume should be no less than the average runoff event during a year (note that this is not the runoff from the average rainstorm). We concur with Grizzard et al. (1986) and recommend that the basin be designed to drain its full water quality volume in no less than 40 hours. In summary, the recommended design parameters for an extended detention basin are as follows:

- Basin volume: somewhat larger than average runoff event; and
- Outlet: size outlet to drain the full basin volume in no less than 40 hours.

23.4.1 Possible Long-term Pollutant Removals

Based on field studies by EPA (1981), Grizzard et al. (1986), Occoquan Watershed Monitoring Laboratory (1986), and Whipple and Hunter (1981), it appears that a properly designed extended detention basin can be expected to achieve the following long-term removal rates:

TSS:	50% to 70%
TP:	10% to 20%
Nitrogen:	10% to 20%
Organic matter:	20% to 40%
Lead:	75% to 90%
Zinc:	30% to 60%
Hydrocarbons:	50% to 70%
Bacteria:	50% to 90%

For planning purposes, it is recommended that the lower end of these ranges be used until local data are available to draw better conclusions.

23.4.2 Example: Sizing Extended Detention Basin

Given: The same 100-acre single family residential development in Arkansas as used in the examples for sizing of ponds with permanent water pool. The area has a runoff coefficient $C_V = 0.25$. The rainfall statistics, as reported by EPA (1986), are as follows:

	Units	Annual Mean	CV
Volume (V)	inches	0.52	1.54
Intensity (I)	in/hr	0.122	1.35

Required: Size an extended detention pond that will remove 50% to 60% of total suspended solids, 40% to 50% of zinc, and 70% lead found in stormwater runoff.

Solution:

1. Calculate the mean runoff event parameters:

   For the sake of this example we are simplifying the assumption and saying that the average runoff volume is estimated here as being two times the runoff from an average storm, namely:

   $$V_R = V\,C_V\,A = \frac{2 \cdot 0.52 \cdot 0.25 \cdot 100 \cdot 43{,}560}{12}$$
   $$= 94{,}400 \text{ cubic feet}$$

2. Thus, the required basin volume:

   $$V_B = V_R = 94{,}400 \text{ cubic feet}$$

3. The outlet for this basin will be a perforated riser. It will overflow only when the volume in step 2 is exceeded. The perforations will be sized to drain the full volume of the pond in no less than 40 hours. They will also assure that the lower one-half of the volume drains in no less than 24 hours. This will insure that the smaller storms will be detained sufficiently to provide pollutant removal.

23.5 RECOMMENDED BASIN CONFIGURATIONS

How water quality (i.e., sedimentation) detention ponds or basins are sized and configured is important in how effectively they will remove pollutants. This was discussed throughout the entire book. However, it is worth going over the key elements again. We again describe the basic elements of good design for both wet ponds and dry detention basins. We also recommend that a coarse pollutant removal basin or forebay be provided for both types of basins.

23.5.1 Removal of Coarse Materials

There are operational advantages in removing coarse material as they enter the storage basin. A forebay near the basin inlet can be built for this purpose. Another approach is to install a separate coarse materials separating facility just ahead of the storage basin. The forebay, or a separate basin, could be lined with concrete or soil cement to facilitate cleaning.

A coarse pollutant basin needs to be sufficiently deep to prevent frequent resuspension of deposited sediments. Also, it needs to have a storage volume sufficient to detain the seasonal average inflow rate for about 5 minutes and a water surface area that provides surface load of approximately 50 feet per hour for a seasonal average hourly inflow rate.

If the runoff from the average storm is expected to be $Q = 0.25 \cdot 100 \cdot 0.122 \cdot 43,560/12 = 11,000$ cubic feet per hour, the average seasonal runoff in this example is arbitrarily assumed to be twice this rate, namely 22,000 cubic feet per hour. Under these conditions, the forebay should have a surface area sized as follows:

$$A = \frac{Q}{V_S}$$

or,

$$A = \frac{22,000}{50} = 440 \text{ square feet}$$

and its volume should be no less than

$$V = (22,000 \text{ ft}^3/\text{hr}) \cdot \frac{5 \text{ min}}{60 \text{ min/hr}} = 1,830 \text{ ft}^3$$

Thus, a basin 4.5 feet deep with a surface area of 15 feet wide and 30 feet long should provide an adequate forebay. All the larger particles should settle out at this location and, as a result, the effective service life of the remaining pond should be extended.

Much of the coarse fraction of the sediment load is not reported in water quality data, since it is carried near the bottom of the drainageway where it is not sampled. Yet, this fraction can account for a large percentage of the solids entering a pond. When these solids settle out inside a pond, they reduce its volume. Thus, cleaning of the forebays when solids accumulate within them is an important part of operating water quality enhancement facilities.

Another feature of a forebay is the installation of a surface skimmer, such as a floating boom between the forebay and the pond, or a fixed skimming baffle. A skimmer will keep most of the floating trash from entering the pond itself and will confine floating debris to the area where it can be removed more easily.

23.5.2 Wet Ponds

In designing a wet pond, namely a basin with a permanent pool of water, the following features are recommended:

- Extend the flow length as much as possible between the inlet and the outlet.
- Minimize the hydraulic surface loading during regularly occurring rainstorms.
- Prevent short circuiting of flow.

- Provide sufficient volume in the permanent pool to capture as much runoff as possible for quiescent sediment removal between storms.
- Enhance conditions for biological treatment between storms.

Let's examine these elements with the aid of Figure 23.9. First, the pond has and elongated shape. Its length to width ratio, if we ignore the forebay area, is equal to three. This provides a fairly long flow path before the water exits at the outlet.

Second, the pond expands gradually from the outlet toward the inlet, insuring that there are no "dead zones." Namely, water entering the pond gradually spreads out and uniformly displaces the water already present in the pond (tries to achieve plug flow).

Third, a baffle is located at the outlet of the forebay. The baffle can be built out of redwood, cedar, or other decay-resistant wood or material and is designed to break up any jets of flow that may have not fully diffused on entering the pond. This type of a device is the best insurance against short-circuiting.

Fourth, size the permanent pool volume in accordance with the recommendations and the procedures described in Section 23.3. To insure that this

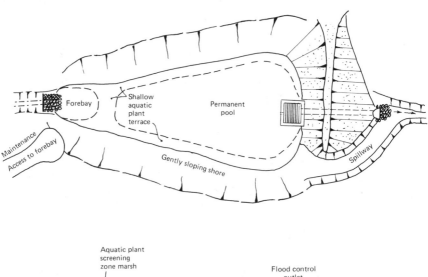

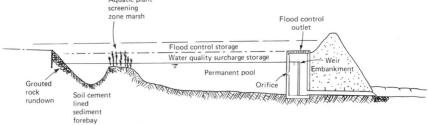

Figure 23.9   Basic configuration of a wet pond.

volume is available for many years of sediment deposition, add approximately 25% more for sediment accumulation.

Fifth, design the outlet so that the average runoff event is captured in a surcharge volume above the permanent pool. This surcharge then is drained off in approximately 12 hours. During larger storms, the excess volume can be allowed to overflow freely at the outlet or at the spillway. This combination of permanent pool and surcharge volume should provide a cost-effective configuration, especially if the water quality pond is a part of a larger basin used to regulate runoff from larger storms such as the 2-, 5-, 10-, or 100-year events.

Sixth, maintain the average pond depth between 4 and 8 feet. Also, provide a 10- to 20-foot wide shallow bench along the shores for safety and to encourage bottom vegetation to develop. It is expected that this vegetation will enhance the biologic treatment characteristics of the pond. When the shoreline has mature bottom vegetation, the pond will also have a more "natural" appearance.

23.5.3 Dry Basins

The configuration of a dry detention basin is, in many respects, similar to what was described for wet ponds. It does, however, need additional features to enhance its use, its aesthetics, and its maintainability. In configuring a dry pond, try to provide the following:

- Extend the flow length between the inlet and the outlet.
- Minimize short circuiting during the filling phase.
- Provide sufficient volume to capture as much runoff as possible for sedimentation to be effective before water leaves the pond.
- Extend its use and aesthetics during periods between storms.
- Provide features to enhance ease of routine maintenance.

Let's examine these with the aid of Figure 23.10. First, the dry basin also has an elongated shape. Its length to width ratio, if we ignore the forebay area, is similar to a wet pond, but this feature is not quite as critical as in a wet pond.

Second, short circuiting is reduced by an outlet that drains very slowly and is packed in coarse gravel. Also, the fill and drain volume (i.e., similar to surcharge volume in a wet pond) needs to be somewhat larger than for a wet pond to trap a wider range of storms.

Third, the basin volume is zoned vertically. The lower level is provided for frequent inundation. It is intended to limit the extent of the muddy or marshy bottom so that the rest of the bottom can be used for passive recreation and can be maintained more easily. If base flows permit, establishing a marsh bottom in the lower zone should also help reduce resuspension of sediment during the filling phase.

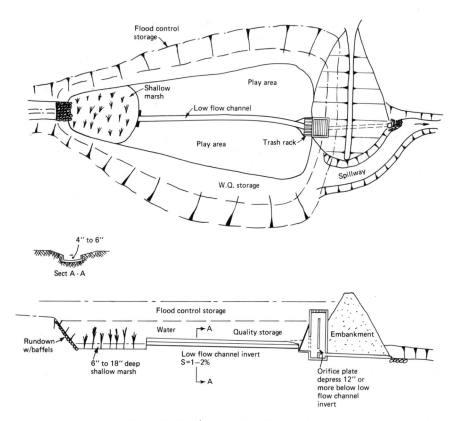

Figure 23.10   Basic configuration of a dry basin.

Fourth, a trickle flow, or a low flow channel, between the inlet and the lower zone will greatly enhance the recreational usability of the upper zone and its maintenance.

REFERENCES

BROWN, C., "Sediment Transportation," *Engineering Hydraulics,* Hunter Rouse, ed., New York, 1950.

CAMP, T. R., "Sedimentation and Design of Settling Tanks," *Transactions of the American Society of Civil Engineers,* Paper No. 2285, pp. 895–958, ASCE, 1946.

DOBBIN, E., "Effect of Turbulence on Sedimentation," *Transactions of the American Society of Civil Engineers,* Paper No. 2218, pp. 629–78, ASCE, 1944.

EPA, *Methodology for Analysis of Detention Basins for Control of Urban Runoff Quality,* U.S. Environmental Protection Agency, EPA440/5-87-001, September, 1986.

GRIZZARD, T. L., RANDALL, C. W., WEAND, B. L., AND ELLIS, K. L., "Effectiveness of Extended Detention Ponds," *Urban Runoff Quality,* American Society of Civil Engineers, 1986.

METROPOLITAN WASHINGTON COUNCIL OF GOVERNMENTS, *Urban Runoff in the Washington Area—Final Report, Washington, D.C. Area Urban Runoff Project,* 1983.

OCCOQUAN WATERSHED MONITORING LABORATORY, *Final Contract Report: Washington Area NURP Project,* Prepared for the Metropolitan Washington Council of Governments, 1986.

RANDALL, C. W., "Stormwater Detention Ponds for Water Quality Control," *Stormwater Detention Facilities—Planning Design Operation and Maintenance,* Proceedings of an Engineering Foundation Conference, ASCE, 1982.

U.S. BUREAU OF RECLAMATION, *Design of Small Dams,* 1973.

WHIPPLE, W., AND HUNTER, J. V., "Settleability of Urban Runoff Pollution," *Journal Water Pollution Control Federation,* pp. 1726–32, 1981.

Index